THE COMPETITION BET
POLYPHOSPHATE-ACCUMULATING ORGANISMS AND
GLYCOGEN-ACCUMULATING ORGANISMS:
TEMPERATURE EFFECTS AND MODELLING

The Competition between Polyphosphate-Accumulating Organisms and Glycogen-Accumulating Organisms: Temperature Effects and Modelling

DISSERTATION

Submitted in fulfilment of the requirements of
the Board for Doctorates of Delft University of Technology
and of the Academic Board of the UNESCO-IHE Institute for
Water Education for the Degree of DOCTOR
to be defended in public
on Monday, June 15, 2009 at 14:00 hours
in Delft, the Netherlands

by

Carlos Manuel LÓPEZ VÁZQUEZ
born in Toluca, México.

Master of Science in Water Sciences
Autonomous University of the State of México, México.

This dissertation has been approved by the supervisors
Prof. dr. ir. M.C.M. van Loosdrecht
Prof. dr. H.J.Gijzen

Members of the Awarding Committee:

Chairman	Rector Magnificus, Delft University of Technology
Prof. dr. R.A.M. Meganck	Vice-Chairman, UNESCO-IHE
Prof. dr. ir. M.C.M. van Loosdrecht	Delft University of Technology, Supervisor
Prof. dr. H.J.Gijzen	UNESCO-IHE, Delft, Supervisor
Prof. dr. D. Brdjanovic	UNESCO-IHE / Delft University of Technology
Prof. J.H.J.M. van der Graaf	Delft University of Technology
Prof. dr. Cheikh Fall	Autonomous University of the State of Mexico, Mexico
Prof. dr. Jürg Keller	The University of Queensland, Australia.

CRC Press/Balkema is an imprint of the Taylor & Francis Group, an informa business

Published by:
CRC Press/Balkema
PO Box 447, 2300 AK Leiden, The Netherlands
e-mail: Pub.NL@taylorandfrancis.com
www.crcpress.com – www.taylorandfrancis.co.uk – www.balkema.nl
ISBN 978-0-415-55896-9 (Taylor & Francis Group)

To my parents
and
Carmen

"Did you ever observe to whom the accidents happen?
Chance favors only the prepared mind".

Louis Pasteur
(1822-1895)

Contents

x

Nomenclature and symbols

The following nomenclature and symbols were used in this dissertation. Notation of parameters related to mathematical modelling is presented in chapter 7.

Abbreviations

ATP	Adenosine triphosphate
ATU	Allyl-N-thiourea
BOD	Biochemical oxygen demand
BPR	Biological phosphorus removal
COD	Chemical oxygen demand
EBPR	Enhanced biological phosphorus removal
FISH	Fluorescence *in situ* hybridization
GAO	Glycogen-accumulating organisms
GLY	Glycogen
HAc	Acetate
HPr	Propionate
HRT	Hydraulic retention time
MLSS	Mixed liquor suspended solids
MLVSS	Mixed liquor volatile suspended solids
N	Nitrogen
NADH	Nicotinamide adenine dinucleotide
OUR	Oxygen uptake rate
P	Phosphorus
PAO	Polyphosphate-accumulating Organisms
PHA	Poly-β-hydroxyalkanoates
PHB	Poly-β-hydroxybutyrate
PHV	Poly-β-hydroxyvalerate
PH2MV	Poly-β-hydroxy-2-methyl-valerate
Poly-P	Poly-phosphate
RBCOD	Readily biodegradable chemical oxygen demand
SBR	Sequencing batch reactor
SBCOD	Slowly biodegradable chemical oxygen demand
SRT	Solids retention time
T	Temperature
VFA	Volatile fatty acids
WWTP	Wastewater treatment plant

Symbols

C-mol	Carbon-mol units
δ	ATP/NADH$_2$ ratio, ATP aerobically produced per NADH$_2$ oxidized
k_{GLY}^{MAX}	Maximum glycogen formation rate
k_{PHA}^{MAX}	Maximum PHA degradation rate
m_{ATP}^{O}	Aerobic ATP maintenance coefficient
m_{OS}	Specific oxygen demand for maintenance
m_S	Specific PHA demand for maintenance
$m_{ATP,GAO}^{An}$	Anaerobic ATP maintenance requirements of GAO
$m_{ATP,PAO}^{An}$	Anaerobic ATP maintenance requirements of PAO
NH$_4^-$-N	nitrogen as ammonium
PO$_4^{3-}$-P	Phosphorus as orthophosphate
$q_{SA,GAO}^{MAX}$	Maximum acetate uptake rate of GAO
$q_{SA,PAO}^{MAX}$	Maximum acetate uptake rate of PAO
$SRT_{MIN,GAO}^{AER}$	Minimum aerobic solid retention time of GAO
$SRT_{MIN,PAO}^{AER}$	Minimum aerobic solid retention time of PAO
r_{GLY}	Specific maximum aerobic glycogen formation rate
r_{O2}	Specific oxygen uptake rate
r_{PHA}	Specific maximum aerobic PHA consumption rate
r_{PHB}	Specific maximum aerobic PHB consumption rate
r_{PHV}	Specific maximum aerobic PHV consumption rate
r_X	Specific maximum biomass production rate
X_{GAO}	Active GAO biomass concentration
X_{PAO}	Active PAO biomass concentration
θ	Temperature coefficient
Y_{ox}^{max}	Maximum yield of biomass production on oxygen
Y_{sx}^{max}	Maximum yield of biomass production on PHA
Y_{ogly}^{max}	Maximum yield of glycogen production on oxygen
Y_{sgly}^{max}	Maximum yield of glycogen production on PHA
Y_{ox}^{max}	Maximum yields of biomass production on oxygen

Summary

The over-enrichment of surface water bodies with nitrogen (N) and phosphorus (P) compounds present in non-treated and treated wastewater, can lead to the excessive growth of algae (e.g. cyanobacteria) and other photosynthetic organisms. This process, known as eutrophication, results in reduced transparency, reduced photosynthetic activity, depletion of oxygen, production of toxic compounds and loss of plant and animal species.

Since phosphorus discharges from municipal wastewater systems are in general higher than from any other point source and phosphorus is the most limiting nutrient for biological growth in natural waters, the operation of efficient phosphorus removal processes in wastewater treatment systems is of major importance to control and prevent eutrophication.

Due to high removal efficiency, economy, environmentally-friendly operation, and potential phosphorus recovery, enhanced biological phosphorus removal (EBPR) in activated sludge processes is a popular and widespread technology in wastewater treatment systems. EBPR can be implemented in activated sludge systems by promoting the enrichment of the system with polyphosphate-accumulating organisms (PAO). This can be achieved by recirculating sludge through anaerobic and anoxic/aerobic conditions, and directing the influent, usually rich in volatile fatty acids (VFA) such as acetate (HAc) and propionate (HPr), to the anaerobic stage.

The main disadvantage of the EBPR process is its apparent instability and unreliability. It is known that EBPR treatment plants may experience process upsets, deterioration in performance and even failure, exceeding the maximum permissible effluent concentrations for phosphorus. Among other factors, the appearance of glycogen-accumulating organisms (GAO), which compete for substrate (VFA) with PAO under anaerobic conditions, has been hypothesised to be the main cause of the deterioration of the EBPR process performance. Therefore, from an EBPR process perspective, GAO are seen as undesirable microorganisms because they compete with PAO for VFA and do not contribute to the biological phosphorus removal.

With a clear need to reduce the phosphorus loads discharged into surface water bodies, and recover and preserve our surrounding ecosystems, it becomes extremely important to find and suggest strategies and control measures to favour the development of PAO and suppress the growth of GAO in EBPR activated sludge systems.

In the present thesis, the effects of key environmental and operating conditions influencing the PAO-GAO competition (such as temperature, types and fractions of volatile fatty acids and pH levels) were addressed from different perspectives through undertaking different studies at both lab- and full-scale and by applying mathematical modelling.

At both short- (Chapters 2 and 3) and long-term temperature effect studies (Chapter 4), GAO (*Competibacter* a known GAO) showed to be able to compete with PAO for substrate only at temperatures higher than 20 $^\circ$C. Below this temperature, GAO did not show metabolic advantages over PAO. Moreover, at temperatures lower than 20 $^\circ$C (e.g. 10 $^\circ$C), the anaerobic metabolism of GAO was inhibited limiting the substrate uptake rate and, therefore, the growth of these microorganisms. In general, GAO had a lower biomass growth rate than PAO, requiring longer SRT, particularly below 20 $^\circ$C. At 10 $^\circ$C, a complete switch in the dominant microbial populations from an enriched GAO to an enriched PAO culture confirmed that GAO cannot adapt to low temperatures.

In full-scale systems (Chapter 5), the occurrence of PAO (*Accumulibacter*) was positively correlated with pH values, suggesting that high pH levels are favorable for these microorganisms at full-scale systems. The appearance of PAO was also influenced by the operation of well-defined denitrification stages that, furthermore, also stimulated the development of denitrifying PAO (DPAO). Quantified GAO populations (*Competibacter*) were only correlated with the organic matter concentrations implying that, according to the conditions of this survey, these microorganisms are only able to grow on the excess of organic matter (or substrate) in the system. *Defluviccocus*-related microorganisms and *Sphingomonas* (presumably GAO) were not observed or only seen in negligible fractions (< 1%) in a few plants indicating that these organisms could be rarely present in municipal EBPR WWTP and, thus, may not be the main competitors of PAO.

In order to monitor the performance of EBPR plants, a practical method was developed using highly enriched PAO (*Accumulibacter*) and GAO (*Competibacter*) cultures which showed to be potentially suitable for the *in situ* quantification of PAO and GAO populations at full-scale systems (Chapter 6). The method relies on the execution of an anaerobic batch test and the measurement of commonly determined parameters such as acetate, orthophosphate and mixed liquor suspended solids.

Using a (mechanistic) mathematical model that incorporated the carbon source, temperature and pH-dependences of PAO (*Accumulibacter*) and GAO (Competibacter and *Alphaproteobacteria*-GAO) the influence of key operating and environmental conditions on the PAO-GAO competition was evaluated (Chapter 7). At low temperature (10 $^\circ$C), PAO were the dominant microorganisms independently upon the pH or carbon source provided. At moderate temperature (20 $^{\circ}$C), the simultaneous presence of acetate and propionate (e.g. 75-25 or 50-50 % acetate to propionate ratios) favoured the growth of PAO over GAO, regardless of the applied pH. However, when either acetate or propionate is supplied as sole carbon source, PAO was only favoured when a high pH (7.5) was applied. At higher temperature (30 $^{\circ}$C), GAO tended to proliferate. Nevertheless, an adequate acetate to propionate ratio (75 – 25 %) and a pH not lower than 7.0 could be used as potential tools to suppress the proliferation of GAO at high temperature (30 $^{\circ}$C).

This research contributed to get a better understanding about the factors affecting the PAO-GAO competition and, thus, the stability and reliability of the EBPR process in activated sludge systems. The findings obtained in this research may prove useful towards optimization of full-scale EBPR plants.

Samenvatting

De verrijking van oppervlaktewater met stikstof en fosfaat aanwezig in onbehandeld en behandeld afvalwater, kan leiden tot overmatige groei van algen (b.v. cyanobacterieën) en andere fototrofe organismen. Dit proces, bekend als eutrofiëring, resulteert in verminderde lichtdoorlaatbaarheid, verminderde fotosyntheseactiviteit, zuurstoftekort, vorming van toxische componenten en verlies van verschillende planten- en diersoorten.

Aangezien fosfaat lozingen van communale afvalwaterzuiveringen over het algemeen groter zijn dan van enige andere puntbron en fosfaat de meest beperkende voedingsstof is voor biologische groei in oppervlaktewater, is het belangrijk om efficiënte fosfaatverwijderingprocessen in afvalwaterzuiveringsinstallaties te hebben ter voorkoming van eutrofiëring.

Wegens het hoge verwijderingrendement en een milieuvriendelijke manier van bedrijven en tevens de mogelijkheid tot fosfaat hergebruik, is biologische fosfaatverwijdering in actief slib processen een populaire en wijd verbreide technologie voor afvalwaterzuiveringssystemen. Biologische defosfatering kan worden geïmplementeerd in actief slibsystemen door middel van het ophopen van polyfosfaat accumulerende organismen (PAO) op basis van hun ecofysiologische karakteristieken. Dit kan worden bereikt door middel van slibrecirculatie door anaërobe en anoxische/aërobe zones, en het leiden van het influent, meestal rijk in vluchtige vetzuren zoals acetaat en propionaat, naar de anaerobe stap.

Het belangrijkste nadeel van het biologische defosfatering is de veronderstelde of schijnbare instabiliteit en onbetrouwbaarheid. Het is bekend dat in biologisch defosfaterende installaties procesverstoringen voorkomen die in het ergste geval leiden tot het niet meer werken van de biologische fosfaatverwijdering, waardoor de maximaal toelaatbare effluentconcentraties voor fosfaat worden overschreden.
Een van de factoren is het ophopen van glycogeen-accumulerende organismen (GAO), die onder anaërobe omstandigheden concurreren om substraat met PAO. Daarom worden, vanuit een EBPR proces perspectief, GAO gezien als ongewenste micro-organismen, omdat ze concurreren met PAO voor vetzuren, maar niet bijdragen aan de biologische fosfaatverwijdering.

Met een duidelijke behoefte om de fosfaatbelasting in oppervlaktewater te verminderen en om onze ecosystemen te behouden, is het zeer belangrijk om strategieën en maatregelen te vinden ter bevordering van de ontwikkeling van PAO in actief slib en om de groei van GAO te onderdrukken.

In dit proefschrift worden de effecten van de belangrijkste milieu- en operationele omstandigheden besproken die van invloed zijn op de PAO-GAO concurrentie (zoals temperatuur, types en fracties van vluchtige vetzuren en pH niveaus). Het onderzoek werd verricht vanuit verschillende perspectieven en middels het uitvoeren van verschillende studies op zowel laboratorium- en praktijkschaal, en ondersteund door wiskundige modellen.

In zowel korte- (hoofdstukken 2 en 3) als lange termijn temperatuur-effect studies (hoofdstuk 4), bleken GAO (m.n. Competibacter een bekende GAO) te kunnen concurreren met PAO voor substraat bij temperaturen hoger dan 20°C. Onder deze temperatuur toonde GAO geen metabole voordelen ten opzichte van PAO. Bij temperaturen lager dan 20° C, werd het anaerobe metabolisme van GAO geremd met beperking van de substraat opnamesnelheid waarmee indirect de groei van deze micro-organismen werd beperkt.

In het algemeen had GAO een lagere biomassaproductie dan PAO, in het bijzonder onder 20°C. Bij 10°C werd een complete omschakeling in de dominante microbiële populaties van een verrijkte GAO naar een verrijkte PAO cultuur bevestigd. GAO kunnen zich blijkbaar niet aanpassen aan lage temperaturen.

In praktijkwaterzuiveringen (hoofdstuk 5), werd het vóórkomen van PAO (Accumulibacter) positief gecorreleerd met hogere pH-waarden, hetgeen suggereerd dat hoge pH-niveaus gunstig zijn voor deze micro-organismen. Het voorkomen van PAO werd ook gunstig beïnvloed door de werking van goed gedefinieerde denitrificatie zones in de waterzuivering, deze zorgden bovendien ook voor de ontwikkeling van denitrificerende PAO (DPAO). Significante GAO populaties (Competibacter) werden alleen gecorreleerd met hoge organische stof concentraties implicerend dat, in de onderzochte systemen, deze micro-organismen zich kunnen handhaven dankzij een overschot aan substraat. Defluviccocus-gerelateerde micro-organismen en Sphingomonas (vermoedelijk GAO) werden niet waargenomen of slechts in verwaarloosbare fracties (<1%) in een paar installaties die aangaven dat deze organismen zelden aanwezig kunnen zijn in de huishoudelijke waterzuivering met EBPR, en dus wellicht niet de belangrijkste concurrenten zijn van PAO.

Een praktische methode werd ontwikkeld met behulp van sterk verrijkte PAO (Accumulibacter) en GAO (Competibacter) cultures die potentieel geschikt is voor de in situ kwantificering van PAO en GAO populaties in praktijksystemen (hoofdstuk 6). De methode berust op de uitvoering van een anaërobe batch test en de meting van parameters zoals acetaat, orthofosfaat en droge stof.

Met behulp van een (mechanistich) wiskundig model dat het effect van de koolstof-bron, de temperatuur en pH-afhankelijkheid van PAO (Accumulibacter) en de GAO (Competibacter en Alphaproteobacteria-GAO) beschrijft, werd de invloed van belangrijke bedrijfs-en omgevingsfactoren op de PAO-GAO competitie geëvalueerd (hoofdstuk 7). Bij lage temperatuur (10°C) zijn PAO de dominante micro-organismen, onafhankelijk van de pH of koolstofbron. Bij gematigde temperatuur (20°C), begunstigd de gelijktijdige aanwezigheid van acetaat en propionaat de groei van PAO t.o.v. GAO, ongeacht de toegepaste pH. Echter, wanneer acetaat of propionaat de enige koolstofbron is, dan zijn PAO alleen in het voordeel bij een hoge pH (7,5). Bij hogere temperatuur (30°C) had GAO vrijwel altijd het voordeel. Toch kan een geschikte acetaat-propionaat ratio (75 - 25%) en een pH niet lager dan 7,0 gebruikt worden als hulpmiddel om de proliferatie van GAO bij hoge temperatuur (30°C) te onderdrukken.

Dit onderzoek heeft bijgedragen tot een beter inzicht in de factoren die de PAO-GAO concurrentie beïnvloeden en zo ook de stabiliteit en de betrouwbaarheid van het proces in EBPR actief slib systemen. De verkregen bevindingen in dit onderzoek zijn nuttig bij het optimaliseren van EBPR installaties zeker in het geval het afvalwater een hogere temperatuur heeft zoals bij de industrie en in tropische gebieden.

1
Chapter

Introduction

Content

1.1. Background

1.1.1. Biological wastewater treatment

In the 19th Century, domestic wastewaters were discharged, through sewers, into rivers and other surface water bodies to convey them to the sea. Rapid urbanization and industrialization exacerbated the problem. The surface water bodies became extremely polluted because the discharged organic compounds served as substrate for bacterial growth, resulting in the utilization and eventual depletion of oxygen, and subsequent death of animal and plant life. As a consequence, the aquatic systems were seriously affected leading to severe environmental issues. Also, due to spreading of pathogens, water borne diseases raised, causing severe public health problems. The first reaction in certain western European countries was the implementation of specific regulations dealing with wastewater treatment and discharge. This action promoted the construction and operation of wastewater treatment plants and the development of wastewater treatment technologies.

Sewage or wastewater treatment is the process of removing contaminants from wastewater by physical, chemical or biological processes. Among them, biological processes, such as aerobic (i.e. activated sludge, trickling filters) and anaerobic systems (i.e. upflow anaerobic sludge blanket reactors, lagoons), were the preferred choice due to their relatively high removal efficiency (which can be even higher than 90 %) and economy compared to physical and chemical processes. Biological treatment systems are based on microbiological processes occurring in nature, but promoting a higher degradation activity and, therefore, removal, by increasing the bacterial concentration. Since wastewater treatment processes depend on the concentration of certain biomass in the systems, the design and operating guidelines of biological wastewater treatment systems have been developed in such a way that they tend to enhance the enrichment of the treatment plants with the desired type and concentration of microorganisms. This manipulation is done because, usually, wastewaters are very diluted and contain neither the required concentration nor the specific types of bacteria needed for the removal of organic and inorganic compounds. Because some of the essential bacteria involved in waste removal are slow growing organisms, the necessary concentrations can often only be reached through the retention of biomass. Meanwhile, the different environments or (biochemical) stages developed within the treatment process are, most of the time, stressing factors whose main objective is to create a metabolic selection to promote the growth and development of the required types of bacteria.

1.1.2. Importance of phosphorus removal for aquatic ecosystems

The first aim of the biological wastewater treatment technology was the removal of biodegradable organic matter, usually quantified in terms of the biochemical or chemical oxygen demand (BOD and COD, respectively). However, the over-enrichment of surface water bodies with nitrogen (N) and phosphorus (P) compounds present in non-treated and treated wastewater, tended to stimulate the growth of algae (e.g. cyanobacteria) and other photosynthetic organisms. This process, known as eutrophication, results in reduced transparency, reduced photosynthetic oxygen generating activity, depletion of oxygen, production of toxic compounds, production of tastes and odours that make the water unsuitable for water supply, loss of plant and animal species, and reduced recreational aesthetics (Gijzen and Mulder, 2001; Seviour *et al.*, 2003; EEA, 2005). Thus, in the last decades, issues like eutrophication and the need to adopt sustainable water and wastewater systems strongly supported the promulgation of specific legislation to limit nutrient discharges (e.g. ammonia and phosphorus) into surface water bodies in Europe (EEC, 1991).

Phosphorus and nitrogen are often the limiting factors for biological growth in natural waters. However, phosphorus is more critical, because cyanobacteria are capable of fixing molecular nitrogen from the atmosphere, thus eliminating the requirement for ammonia (NH_3-N) or nitrate (NO_3^--N) (Seviour *et al.*, 2003). Experiments with large water reservoirs have shown that no eutrophication occurs when the phosphorus concentration is reduced to 8-10 µg P/L, even at nitrogen concentrations as high as 4-5 mg N/L (Korstee *et al.*, 1994).

Phosphorus is an essential (macro) nutrient to the growth of biological organisms (Metcalf and Eddy, 2003). Since nearly every cellular process that uses energy obtains it in the form of adenosine-5'-triphosphate (ATP), it is a key element in all known forms of life. Moreover, phosphorus is also a vital component of biological molecules such as deoxyribonucleic acid (DNA), ribonucleic acid (RNA) and phospholipids in the cell membrane.

In Europe, nutrient discharges from municipal wastewater systems are in general higher than from any other point source, accounting for more than 70% of the total phosphorus discharged into surface water bodies through the sewerage (Table 1.1). Other activities, like agriculture, industry and aquaculture are responsible for the rest of the phosphorus load released to the environment. This points out wastewater treatment plants as key players concerning the major point sources for phosphorus (Seviour *et al.*, 2003) and underlines the importance of the operation of efficient phosphorus removal processes in these treatment systems. Therefore, in practice, eutrophication and, consequently, the excessive algal bloom, could be controlled and prevented through reducing the maximum permissible phosphorus concentration discharged into the surface water bodies by wastewater treatment plants. Within this context, the effluent discharge requirements range from 1 to 2 mg/L of total phosphorus depending on plant size, location and potential impact on receiving waters (EEC, 1991; Metcalf and Eddy, 2003). In contrast, the effluent requirement for total nitrogen ranges from 10 to 15 mg/L.

Table 1.1. Percentage split of point source discharges to the Baltic, North Sea and Danube Basin (source: EEA, 2005).

Point source discharges of phosphorus to the Baltic (2000)		
Point source	Inland waters	Direct
Municipal wastewater	85	81
Industry	14	14
Fish farms	1	4
Point source discharges of phosphorus to the North Sea (2000)		
Sewage treatment works	75	68
Households not connected to sewage treatment works	10	15
Industry	14	16
Aquaculture	1	1
Point source discharges of phosphorus to the Danube Basin (1996-1997)		
Municipal	73	78
Industrial	19	15
Agricultural	8	7

1.2. Enhanced biological phosphorus removal (EBPR) in activated sludge systems

1.2.1. Activated sludge wastewater treatment systems

The activated sludge process represents the most widespread technology for wastewater treatment and is a component of the largest biotechnology industry in the world (Seviour and Blackall, 1999). Because of its reliability, flexibility and adaptability it could be considered the most common and popular wastewater treatment process.

Presently, several configurations for the activated sludge process have been developed. However, all of them consist, at least, of two basic stages: an aeration tank and a secondary clarifier. The two phases can be placed in separate units (conventional activated sludge system) or in the same unit like in sequencing batch reactors (SBR) (Wilderer *et al.*, 2001). In the aeration tank, mixed cultures, which form the activated sludge, grow in suspension as three-dimensional aggregated communities called flocs. In the secondary clarifier, quiescent conditions are needed to allow the settling and further removal of the flocs, in order to produce an effluent low in suspended solids. Part of the settled solids is recycled from the secondary clarifier to the aerobic tank (return activated sludge) aiming at keeping certain biomass concentration in the aerated stage, meanwhile, excess of solids is wasted (waste activated sludge).

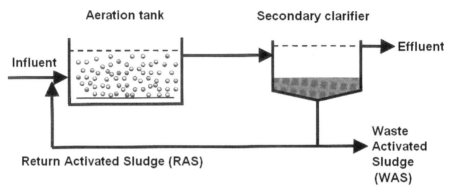

Figure 1.1. Simplified scheme of an activated sludge wastewater treatment system.

1.2.2. Enhanced biological phosphorus removal (EBPR)

Due to its popularity and widespread application around the world, the activated sludge process was one of the preferred treatment systems for the implementation of phosphorus removal technologies by biological or chemical precipitation means (using calcium, aluminum or iron salt addition) or through combined biological removal/chemical precipitation systems.

In particular, enhanced biological phosphorus removal (EBPR) can be relatively easy implemented in activated sludge wastewater treatment systems. EBPR is achieved by recirculating sludge through anaerobic and anoxic/aerobic conditions, and directing the influent, usually rich in volatile fatty acids (VFA) such as acetate (HAc) and propionate (HPr), to the anaerobic stage (Figure 1.2). This promotes the enrichment of the system with polyphosphate-accumulating organisms (PAO) (Mino *et al.*, 1998).

In the anaerobic stage, PAO store the VFA present in the influent as poly-β-hydroxyalkanoates (PHA). Depending on the type of VFA supplied (e.g. HAc or HPr), the PHA stored under anaerobic conditions could be comprised by poly-β-hydroxybutyrate (PHB), poly-β-hydroxyvalerate (PHB) or poly-β-hydroxy-2-methylvalerate (PH$_2$MV) (Smolders *et al.*,

1994a; Oehmen *et al.*, 2005c). Most of the energy required for the transport of VFA through the cell membrane and further storage as PHA is provided by the hydrolysis of internally stored poly-phosphate (poly-P) (Mino *et al.*, 1998). Meanwhile, it is assumed that glycogen provides the reducing power necessary for PHA storage (Smolders *et al.*, 1994a; Mino *et al.*, 1998). poly-P hydrolysis also supplies the energy required to cover the anaerobic maintenance requirements (Smolders *et al.*, 1994a). As consequence of poly-P hydrolysis, orthophosphate (PO_4^{3-}-P) is released into the bulk liquid under anaerobic conditions (Figure 1.2).

In the aerobic (or anoxic) phase, PAO utilize the PHA stored under anaerobic conditions as carbon and energy source to take up higher amounts of orthophosphate than those released in the anaerobic stage (Smolders *et al.*, 1994b). The intracellular stored PHA is also used for (Smolders *et al.*, 1995): (i) the replenishment of the intracellular glycogen pool, (ii) biomass growth, and (iii) to cover the aerobic maintenance needs. Net P-removal from the wastewater is achieved through removal of waste activated sludge (WAS) in the end or after the aerobic phase, when sludge contains a high poly-P content (Figure 1.2).

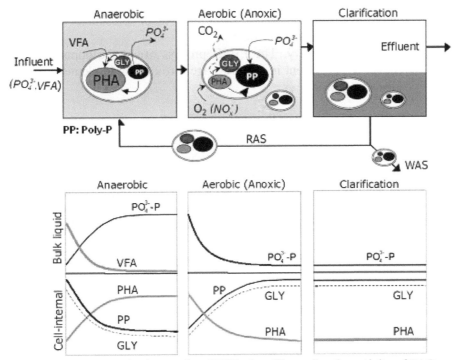

Figure 1.2. Conceptual scheme of an EBPR plant illustrating the activity of PAO.

1.2.3. Advantages of the EBPR process

When operated successfully, the EBPR process is a relatively inexpensive and environmentally sustainable option for P-removal (Oehmen *et al.*, 2007). The advantages of EBPR compared to chemical P-removal are (Korstee *et al.*, 1994; van Loosdrecht *et al.*, 1997; Jansen *et al.*, 2002; Metcalf and Eddy, 2003; Barat and van Loosdrecht, 2006):

    a. High removal efficiency. EBPR is capable of achieving P effluent concentrations of less than 1 mg/L, whereas, to achieve a similar effluent concentration, chemical P-removal may usually require high chemical dosages depending upon wastewater characteristics and operating conditions.

    b. No increase on sludge production. Due to the absence of chemicals for P precipitation, there is not generation of chemical sludge that may increase the total sludge production.

    c. Economic. Since there is not an increase on the total sludge production the sludge treatment, disposal and management costs are lower. Moreover, chemical sludge may be more difficult to dewater, increasing the dewatering costs.

    d. Environmentally-friendly. No downstream ecological effects due to the absence of aluminum, iron and other presumably harmful residuals in the treated effluent.

    e. Potential P-recovery. The EBPR process offers the opportunity to recover the P present in the sewerage through the implementation of side-stream processes.

1.2.4. EBPR process upsets and deterioration

The main disadvantage of the EBPR process is its apparent instability and unreliability. It is known that EBPR plants may experience process upsets, deterioration in performance and even failure, exceeding the maximum permissible effluent concentrations (Seviour *et al.*, 2003; Thomas *et al.*, 2003; Oehmen *et al.*, 2007). According to diverse reports in literature, EBPR is affected by a number of environmental and operating factors.

Excessive aeration after heavy rainfalls and weekends may decrease the level of intracellular stored polymers (PHA and glycogen), affecting the phosphorus removal performance (Brdjanovic *et al.*, 1998b; Lopez *et al.*, 2006). The intrusion of nitrate or nitrite (NO_3^--N and NO_2^--N, respectively) in the anaerobic zone has also been associated to the deterioration of the EBPR activity. The simultaneous presence of NO_3^--N or NO_2^--N and VFA induces anoxic conditions, favouring the consumption of VFA for denitrification processes by both ordinary heterothoph organisms (OHO) and PAO (Kuba *et al.*, 1994; Puig *et al.*, 2007), decreasing the phosphorus removal efficiency. Moreover, van Niel *et al.* (1998) and Saito *et al.* (2004) described the inhibitory effect that NO_3^--N and NO_2^--N could have on the EBPR process. The proper installation and operation of on-line sensors coupled to the aeration systems and internal recirculation rates has helped to keep better operating conditions avoiding excessive aeration periods and reducing the intrusion of NO_3^--N and NO_2^--N (Jansen *et al.*, 2002; Olsson, 2006).

Regarding the applied anaerobic hydraulic retention time (HRT), Matsuo (1994) observed an improvement on the P-removal performance after the anaerobic HRT was extended. At full-scale systems (Jansen *et al.*, 2002), limited P-removal capacity is usually observed in those plants where a short anaerobic HRT is applied (e.g. shorter than 0.50 hours), particularly during winter. This underlines the importance of the length of the anaerobic stage on the EBPR process. Contrary to the growth of OHO, the EBPR is not limited by the total solid or sludge retention time (SRT), but by the length of the aerobic SRT (Brdjanovic *et al.*, 1998c), which should be long enough to oxidize the PHA stored in the anaerobic stage. By incorporating the temperature dependencies of PAO reported in literature (Brdjanovic *et al.*, 1998a; Henze *et al.*, 2000; Jansen *et al.*, 2002) on the design guidelines for EBPR wastewater treatment plants, these operational limitations have been reduced.

In other cases, the appearance of glycogen-accumulating organisms (GAO), which compete for VFA with PAO, has been hypothesised to be the main cause of the deterioration of the EBPR process performance (Cech *et al.*, 1993; Satoh *et al.*, 1994; Liu *et al.*, 1994, 1996; Filipe *et al.*, 2001a; Thomas *et al.*, 2001; Oehmen *et al.*, 2007). However, to present, the mechanisms that influence the occurrence of GAO have not been fully addressed.

1.3. Glycogen accumulating organisms (GAO)

GAO, likewise PAO, are also able to store VFA as PHA under anaerobic conditions. However, since they do not store poly-P, intracellular stored glycogen is used as both energy and carbon source for VFA uptake without exhibiting the typical anaerobic P-release and subsequent aerobic P-uptake from PAO (Mino *et al.*, 1998; Filipe *et al.*, 2001a; Zeng *et al.*, 2002, 2003a; Oehmen *et al.*, 2006b) (Figure 1.3). In the aerobic phase (Zeng *et al.*, 2003a; Oehmen *et al.*, 2006b), GAO utilize the PHA previously stored in the anaerobic stage for: (i) replenishment of glycogen, (ii) biomass growth, and (iii) to cover the aerobic maintenance requirements (Figure 1.3). At lab- and full-scale plants, different reports have described how the presence and proliferation of GAO led to sub-optimal and deteriorated EBPR performance

(Saunders *et al.*, 2003; Thomas *et al.*, 2003; Gu *et al.*, 2005) and, in extreme cases, failure (Satoh *et al.*, 1994; Filipe *et al.*, 2001a). Therefore, from an EBPR process perspective, GAO are seen as undesirable microorganisms because they compete with PAO for VFA and do not contribute to the biological phosphorus removal.

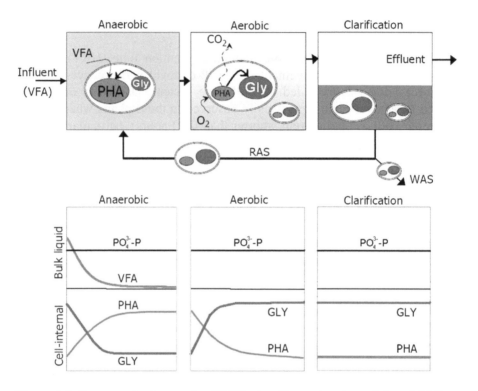

Figure 1.3. Conceptual scheme of an EBPR plant illustrating the activity of GAO.

Minimizing the growth of GAO in EBPR plants could increase the cost-effectiveness of the process; however, assessing the microbial dynamics that takes place in such systems (e.g. the PAO-GAO competition) has been a recent development that is still expanding (Oehmen *et al.*, 2007). Further research is required to increase our knowledge about which and how different environmental and operating factors affect the metabolisms of PAO and GAO in order to get a better understanding about the PAO-GAO competition. This could provide important information to comprehend the EBPR process performance under different conditions, which may lead to the proposal of control measures to improve its stability and reliability.

1.4. Microbial identification

1.4.1. Identification of PAO

In spite of the difficulty in the isolation of PAO (Mino *et al.*, 1998; Seviour *et al.*, 2003; Oehmen *et al.*, 2007), this group of microorganisms has been identified in EBPR systems with the help of molecular techniques such as fluorescence in situ hybridization (FISH), 16S rRNA-based clone libraries or denaturing gradient gel electrophoresis (DGGE) (Seviour *et al.*, 2003). Using FISH (Amann, 1995), Wagner *et al.* (1994) showed that the originally proposed PAO, *Acinetobacter*, had little significance in full-scale EBPR plants. Applying the same molecular technique, Bond *et al.* (1999) observed that organisms from the subclass 2 *Betaproteobacteria* closely related to *Rhodocyclus* were dominant in EBPR systems exhibiting a good P-removal. Hesselman *et al.* (1999) named these microorganisms as "*Candidatus Accumulibacter Phosphatis*", often abbreviated to *Accumulibacter*. In different lab- and full-scale studies, *Accumulibacter* exhibited the characteristic PAO phenotype: anaerobic/aerobic cycling of poly-P and PHA coupled to anaerobic P-release and aerobic P-uptake, supporting the hypothesis that the *Accumulibacter* group are PAO (Hesselman *et al.*, 1999; Crocetti *et al.*, 2000). Different FISH probes for *Accumulibacter* that target the organisms at different areas of the 16S rRNA have been developed (Table 1.2), becoming a useful tool for the identification and quantification of *Accumulibacter*.

While other studies suggest that there may be various groups of PAO (Zilles *et al.*, 2002; Wong *et al.*, 2005), research carried out at lab- and full-scale, regarding physiological and biomass activity tests, support the hypothesis that *Accumulibacter* appears to be the only known PAO group (Oehmen *et al.*, 2007). Kong *et al.* (2005) showed that two morphotypes of *Actinobacteria* (related to *Tetrasphera*) were abundant in different EBPR plants, being able to remove phosphorus aerobically after consuming aminoacids under anaerobic conditions. However, due to a lack of data concerning their biochemical mechanisms and biomass activity, the contribution of these microorganisms to the EBPR process performance has not been defined.

Table 1.2. 16S rRNA-targeted probes used for FISH detection of (potential) PAO (Oehmen *et al.*, 2007).

Probe	Sequence 5'-3'	Specificity	Reference
PAO462	CCGTCATCTACWCAGGGTATTAAC	Most *Accumulibacter*	Crocetti *et al.* (2000)
PAO651	CCCTCTGCCAAACTCCAG	Most *Accumulibacter*	Crocetti *et al.* (2000
PAO846	GTTAGCTACGGCACTAAAAGG	Most *Accumulibacter*	Crocetti *et al.* (2000
RHC439	CNATTTCTTCCCCGCCGA	*Rhodocyclus* and *Accumulibacter*	Hesselmann *et al.* (1999)
RHC175	TGCTCACAGAATATGCGG	Most *Rhodocyclaceae*	Hesselmann *et al.* (1999)
PAO462b	CCGTCATCTRCWCAGGGTATTAAC	Most *Accumulibacter*	Zilles *et al.* (2002a)
PAO846b	GTTAGCTACGGYACTAAAAGG	Most *Accumulibacter*	Zilles *et al.* (2002a)
Actino-221[a]	CGCAGGTCCATCCCAGAC	*Actinobacteria* (potential PAO)	Kong *et al.* (2005)
Actino-658[a]	TCCGGTCTCCCCTACCAT	*Actinobacteria* (potential PAO)	Kong *et al.* (2005)

[a]. Requires competitor or helper probes.

1.4.2. Identification of GAO

The term GAO defines the phenotype of microorganisms that store glycogen aerobically and consume it anaerobically as their main carbon and energy source for taking up carbon sources and store them as PHA (Mino *et al.*, 1995; Oehmen *et al.*, 2007). Similar to *Accumulibacter*, several difficulties have been faced up when trying to isolate GAO (Mino *et al.*, 1998; Oehmen *et al.*, 2007). Nevertheless, through applying DGGE of PCR-amplified methods and developing 16S r-RNA clone libraries, FISH probes to target the dominant microorganisms from a deteriorated EBPR sludge were designed (Table 1.3) (Nielsen *et al.*, 1999; Crocetti *et al.* 2002). These microorganisms, which belonged to the *Grammaproteobacteria*, showed the typical GAO phenotype (Nielsen *et al.*, 1999; Crocetti *et al.*, 2002). Based on an EBPR process perspective, Crocetti *et al.* (2002) named these organisms as "*Candidatus Competibacter Phosphatis*" often abbreviated as *Competibacter*, or also known as the GB lineage. Kong *et al.* (2002) designed additional FISH probes to target a novel group of GAO, which was observed in lab- and full-scale studies, providing a more robust molecular tool for the identification of *Competibacter* (Table 2).

In recent years, other types of GAO have been observed mainly in lab-scale but also at full-scale studies, although usually in low abundance (Beer *et al.*, 2004; Wong *et al.*, 2004; Meyer *et al.*, 2006; Burow *et al.*, 2007). Beer *et al.* (2004) designed FISH probes to target a group of bacteria related to the *Sphingomonadales*, from *Alphaproteobacteria*, which had been observed in a deteriorated EBPR lab-scale reactor (Table 2). Using rRNA-based stable isotope probing followed by full-cycle rRNA analysis, Meyer *et al.* (2006) identified two clusters of *Alphaproteobacteria* that performed the typical GAO phenotype in a HPr-fed lab-scale reactor. Interestingly, Cluster 1 was similar to that observed by Wong *et al.* (2004) while two new FISH probes were designed to target cluster 2 (Meyer *et al.*, 2006). However, there appear to exist other GAO belonging to the *Alphaproteobacteria* group as Oehmen *et al.* (2006b) observed that not all the microorganisms performing the GAO phenotype bind the currently available FISH probes.

Table 1.3. 16S rRNA-targeted probes used for FISH detection of (potential) GAO (Oehmen *et al.*, 2007).

Probe	Sequence 5'-3'	Specificity	Reference
Gam1019	GGTTCCTTGCGGCACCTC	Some *Gammaproteobacteria*	Nielsen *et al.* (1999)
Gam1278	ACGAGCGGCTTTTTGGGA	Some *Gammaproteobacteria*	Nielsen *et al.* (1999)
GAOQ431	TCCCCGCCTAAAGGGCTT	Some *Competibacter*	Crocetti *et al.* (2002)
GAOQ989	TTCCCCGGATGTCAAGGC	Some *Competibacter*	Crocetti *et al.* (2002)
GB	CGATCCTCTAGCCCACT	*Competibacter* (GB group)	Kong *et al.* (2002b)
GB_G1[a] (GAOQ989)	TTCCCCGGATGTCAAGGC	Some *Competibacter*	Kong *et al.* (2002b)
GB_G2[a]	TTCCCCAGATGTCAAGGC	Some *Competibacter*	Kong *et al.* (2002b)
GB_1 and 2	GGCTGACTGACCCATCC	Some *Competibacter*	Kong *et al.* (2002b)
GB_2	GGCATCGCTGCCCTCGTT	Some *Competibacter*	Kong *et al.* (2002b)
GB_3	CCACTCAAGTCCAGCCGT	Some *Competibacter*	Kong *et al.* (2002b)
GB_4[a]	GGCTCCTTGCGGCACCGT	Some *Competibacter*	Kong *et al.* (2002b)
GB_5	CTAGGCGCCGAAGCGCCC	Some *Competibacter*	Kong *et al.* (2002b)
GB_6 (Gam1019)	GGTTCCTTGCGGCACCTC	Some *Competibacter*	Kong *et al.* (2002b)
GB_7[a]	CATCTCTGGACATTCCCC	Some *Competibacter*	Kong *et al.* (2002b)
SBR9-1a	AAGCGCAAGTTCCCAGGTTG	*Sphingomonas*-related organisms (potential GAO)	Beer *et al.* (2004)
SBR9-1b	TGTTAGGGGCTTAGACCT	*Sphingomonas*-related organisms (potential GAO)	Beer *et al.* (2004)
SBR8-4	CACCGAAGCACTAAGTGCCC	*Sphingomonas*-related organisms (potential GAO)	Beer *et al.* (2004)
TFO_DF218	GAAGCCTTTGCCCCTCAG	*Defluviicoccus*-related organisms (cluster 1)	Wong *et al.* (2004)
TFO_DF618	GCCTCACTTGTCTAACCG	*Defluviicoccus*-related organisms (cluster 1)	Wong *et al.* (2004)
DF988[a]	GATACGACGCCCATGTCAAGGG	*Defluviicoccus*-related organisms (cluster 2)	Meyer *et al.* (2006)
DF1020[a]	CCGGCCGAACCGACTCCC	*Defluviicoccus*-related organisms (cluster 2)	Meyer *et al.* (2006)

[a]. Requires competitor or helper probes.

1.4.3. Other methods for the identification of PAO and GAO

FISH (Amann, 1995) has become a standardized and reliable technique not only for microorganism identification but also to estimate the biomass fractions. However, this technique requires sophisticated and costly equipment (e.g. a specialized equipped microscope), which limits a wider use. Brdjanovic et al. (1999) proposed a bioassay to estimate the GAO to PAO ratio in activated sludge. The method was based on HAc and PO_4^{3-}-P measurements, with and without depletion of the poly-P pool, but it was not experimentally validated on activated sludge. Later, Filipe et al. (2001d) and Schuler and Jenkins (2003) proposed other simple methods to estimate the PAO and GAO dominance, based on the observed EBPR activity. Filipe et al. (2001d) proposed to evaluate the possible presence of GAO, by executing two anaerobic batch tests at different pH values (6.5 and 8.0) and measuring the consumed glycogen and accumulated PHA. Schuler and Jenkins (2003) suggested using either the anaerobic P released to HAc consumption ratio, or the glycogen degradation to HAc consumption ratio. However, these assays provide a means of determining the relative PAO and GAO populations or activities, but they do not, by themselves, provide an estimate of the PAO and GAO fractions of the total biomass. Moreover, the applicability of these methods could be limited, since, PHA and glycogen are parameters difficult to determine *in situ* at full-scale wastewater treatment plants.

As hypothesized by Brdjanovic and colleagues (1999), a simple and practical method to quantify the PAO and GAO populations at lab- and full-scale systems, only based on commonly measured and reliable analytical parameters (such as HAc and PO_4^{3-}-P) could be an useful and fast tool to control and monitor the EBPR process performance, even in small wastewater treatment plant facilities.

1.5. Factors affecting the PAO-GAO competition

Diverse studies have been undertaken aiming at getting a better understanding about the influence of different environmental and operating conditions on the PAO-GAO competition. The effects of temperature, types of influent carbon sources, pH and influent P/VFA ratio, among other parameters, have been observed to play an important role on the competition between PAO and GAO.

1.5.1. Temperature effects

Most of the lab-scale studies carried out to address the effects of temperature on the PAO-GAO competition agree on the statement that, at wastewater temperatures higher than 20 °C, the activity of the BPR process tends to deteriorate and GAO become the dominant microorganisms (Panswad *et al.*, 2003; Erdal *et al.*, 2003; Whang and Park, 2002, 2006). However, the underlying mechanisms of the EBPR process deterioration and actual temperature effects on the metabolism of PAO and GAO remain unclear since all those studies were not performed using enriched PAO and GAO cultures. At full-scale systems, different studies have described the dominance of GAO and the EBPR performance deterioration of wastewater treatment plants handling warm effluents (where sewage temperature is higher than 20 °C) (Grady and Filipe, 2000; Park *et al.*, 2001; Jobàggy *et al.*, 2002; Gu *et al.*, 2005; Barnard and Steichen, 2006). These corroborate the conclusions withdrawn from lab-scale studies.

Brdjanovic *et al.* (1997, 1998a) carried out a systematic study on an enriched PAO culture in order to understand the short- and long-term temperature effects on the EBPR process. On the contrary, analogous systematic studies with an enriched GAO culture have not been reported yet. Since PAO and GAO compete for substrate under anaerobic conditions, the effects of temperature on their anaerobic metabolisms play a crucial role. Moreover, despite the fact that biomass production and glycogen storage take place under aerobic conditions, limited attention has been paid to the effects of temperature on the aerobic metabolism of GAO. A systematic study on an enriched GAO culture could provide important information to understand the occurrence of these microorganisms at full-scale wastewater treatment plants (WWTP). Furthermore, the anaerobic and aerobic temperature dependencies of GAO could be combined to model the interaction between PAO and GAO at different temperatures, which may furthermore help to comprehend the stability of the EBPR process at different weather conditions. Therefore, there is a clear need for studying and determining the temperature dependencies of the metabolism of GAO.

1.5.2. The effect of the carbon sources

Distinct carbon sources, mostly VFA but also non-VFA, such as glucose and ethanol, have been observed to have a strong influence on the EBPR microbial communities. Since HAc and HPr are the dominant carbon sources present in the influent of full-scale treatment plants (Mino *et al.*, 1998; Meijer *et al.*, 2002; Oehmen *et al.*, 2007), most of the research has focused on the effect of VFA on the PAO-GAO competition and, therefore, on the EBPR stability. Either stable (Kuba *et al.*, 1994; Smolders *et al.*, 1995; Brdjanovic *et al.*, 1998a) or unstable EBPR processes (Filipe *et al.*, 2001a; Satoh *et al.*, 1994; Oehmen *et al.*, 2007) have been reported when using HAc as sole carbon source. Whereas, stable EBPR systems appear to be achieved when HPr is supplied (Chen *et al.*, 2004; Pijuan *et al.*, 2004; Oehmen *et al.*, 2006a, 2007). However, when studying the VFA effects on the metabolisms of the EBPR microbial communities, the use of either HAc or HPr as sole carbon source does not seem to ensure the dominance of PAO. According to Oehmen *et al.* (2005b,c, 2006a), *Accumulibacter* are able to take up HAc and HPr with the same efficiency and at a similar kinetic rate (around 0.20 C-mol/C-mol/h). Meanwhile, the currently known GAO (*Competibacter* and *Alphaproteobacteria-GAO*) have different carbon source preferences. While *Competibacter* can take up HAc at the same rate like *Accumulibacter* (at about 0.20 C-mol/C-mol/h), their HPr uptake is practically negligible (Oehmen *et al.*, 2005b, 2006a). On the other hand, *Alphaproteobacteria-GAO* can compete with *Accumulibacter* for HPr because their HPr uptake rate is similar, but they are not able to compete for HAc because they take up HAc at a lower rate (approximately 50 %) than PAO (Oehmen *et al.*, 2005b, 2006a,b; Dai *et al.*, 2007). The high preference of *Accumulibacter* for both HAc and HPr led to the development of a control strategy, which consists of periodically alternate the carbon feed between HAc and HPr, to minimize the growth of GAO (Lu *et al.*, 2006). Despite that this strategy seems to be promising, to periodically alternate these two VFA may face up operating limitations at full-scale systems. Nevertheless, taking into account that the known GAO strains are not able to take up HAc and HPr as efficiently as PAO, it appears that in order to suppress the proliferation of GAO these VFA should be supplied following certain HAc to HPr ratios. Moreover, in full-scale EBPR systems, Thomas *et al.* (2003) and Zeng *et al.* (2006) observed that the HAc to HPr ratio can be controlled through adjusting the operational conditions of the prefermenters. Thus, to define a proper HAc to HPr ratio to favour PAO over GAO could lead to more stable and reliable EBPR processes at both lab- and full-scale systems.

1.5.3. The effect of pH

Several studies have postulated that a pH higher than 7.25 is necessary to keep a good EBPR process performance (Filipe *et al.*, 2001c; Schuler and Jenkins, 2002; Oehmen *et al.*, 2005a). The main reason appears to be that, assuming that the intracellular pH is kept constant, an increase in the pH in the water phase creates a higher pH gradient and an increase in the electrical potential difference across the cell membrane (Smolders *et al.*, 1994a; Filipe *et al.*, 2001a, 2001d). This results in a higher energy requirement for HAc transport through the cell membrane and maintenance at higher pH levels. In the case of PAO, the higher energy needs lead to higher poly-P degradation and, consequently, to a higher anaerobic P-release (Smolders *et al.*, 1994a). The rest of the anaerobic metabolic processes of PAO, including their HAc uptake rate, seem to be independent upon pH changes (Smolders *et al.*, 1994a; Filipe *et al.*, 2001d). To the contrary, GAO lose competitive advantages over PAO since their HAc uptake rate decreases as pH rises (Filipe *et al.*, 2001a). A possible explanation could be that PAO rely on two energy sources (poly-P and glycogen), whereas GAO only on the intracellular glycogen. Thus, as the pH level increases, the anaerobic glycolysis seems to be affected and becomes unable to provide the energy required to cover both the energy necessary for HAc uptake and anaerobic maintenance (Filipe *et al.*, 2001a).

Smolders *et al.* (1994a) and Filipe *et al.* (2001a, 2001d) proposed linear expressions concerning the pH dependency of the anaerobic stoichiometry of PAO and GAO for the case of HAc through the α parameter, which represents the energy (ATP) necessary for the transport of substrate over the cell membrane. Furthermore, Filipe *et al.* (2001a) developed a Monod-type expression to describe the pH-effect on the anaerobic substrate uptake rates of GAO. Regarding the PAO and GAO cultures cultivated on HPr, Oehmen *et al.* (2005c) observed a non-linear pH-dependency of the anaerobic stoichiometry of PAO, meanwhile the anaerobic metabolism of GAO appeared to be not affected by pH fluctuations (Oehmen *et al.*, 2006b).

1.5.4. Other factors affecting the PAO-GAO competition

While most of the research regarding the PAO-GAO competition has been mainly focused on the effect of carbon source, pH and temperature, other factors, such as the influent P/VFA ratio (Liu *et al.*, 1997; Mino *et al.*, 1998; Schuler and Jenkins, 2003), SRT (Rodrigo *et al.* 1999; Whang and Park, 2006; Whang *et al.*, 2007) and the origin of the inoculum to seed lab-scale EBPR systems (Matsuo *et al.*, 1982; Satoh *et al.*, 1994; Mino *et al.*, 1998), have been also hypothesized to affect the competition between PAO and GAO.

The influent P/VFA ratio is considered to be an important parameter that affects the PAO-GAO competition (Liu *et al.*, 1997; Mino *et al.*, 1998; Schuler and Jenkins, 2003). All those studies agree that a high influent P/VFA ratio (above 0.12 P-mol/C-mol) tends to favour the activity of PAO; whereas, influent P/VFA ratios lower than 0.02 P-mol/C-mol create P limiting conditions that, due to a lack of P, suppress the growth of PAO, resulting beneficial for GAO. Thus, in lab-scale studies, enriched PAO cultures have been cultivated by using high P/VFA ratios (Liu *et al.*, 1997; Schuler and Jenkins, 2003); meanwhile, enriched GAO cultures have been obtained through applying low influent P/VFA ratios (Zeng *et al.* 2003a; Oehmen *et al.*, 2006b; Dai *et al.*, 2007). Nevertheless, enriched PAO cultures have been also achieved at (relatively) low P/VFA ratios of about 0.04 P-mol/C-mol (Smolders *et al.*, 1995; Kuba *et al.*, 1996; Brdjanovic *et al.*, 1998a) and GAO at P/VFA ratios higher than 0.10 P-mol/C-mol (Satoh *et al.*, 1994; Filipe *et al.*, 2001a). These observations imply that the influence of other factors, besides the influent P/VFA ratio, may have a higher effect on the PAO-GAO competition.

Concerning the SRT-effects, Rodrigo *et al.* (1999) concluded that shorter SRT are beneficial for PAO after observing that the EBPR biomass activity decreased as the SRT was extended, suggesting that GAO may tend to dominate at longer SRT. In an acetate-fed lab-scale reactor operated at 30 °C and pH 7.5, Whang and Park (2006) observed the switch in the dominant microbial population from an enriched-GAO to an enriched-PAO culture when lowering the applied SRT from 10 to 3 d. Whang *et al.* (2007), through a model-based analysis, inferred that, under the operating conditions applied by Whang and Park (2006), GAO had a lower net biomass growth rate than PAO and, therefore, were outcompeted after the SRT was shortened. However, those studies do not provide further details related to the effect of the operating and environmental conditions in order to get a better understanding about the involved microbial mechanisms. Moreover, considering the temperature applied by Whang and Park (2006) of 30 °C, their observations tend to partially contradict previous hypotheses regarding the effect of higher temperatures on the PAO-GAO competition. Although studies regarding the minimum anaerobic and aerobic SRT of PAO are available in literature (Mamais and Jenkins, 1992; Matsuo, 1994; Brdjanovic *et al.*, 1998c), no data concerning the effect of SRT on GAO cultures have been reported. To evaluate the SRT effects on GAO, which, like those of PAO, may be highly dependent upon temperature (Brdjanovic *et al.*, 1998c), could provide an important insight not only into the influence of the SRT but also into the net effects of temperature on the PAO-GAO competition.

The influence of the inoculum to seed lab-scale EBPR systems appears to play also a role on the competition between PAO and GAO. Matsuo *et al.* (1982), cited by Satoh *et al.* (1994) and Mino *et al.* (1998), started up two parallel SBR and, apparently due to the origin of the seeded sludge, one succeed while the other failed to achieve EBPR activity. In another study, Filipe *et al.* (2001a) started up an EBPR SBR with activated sludge from a municipal plant. Whereas in similar experiments enriched PAO cultures were cultivated through applying practically identical operating and environmental conditions (Smolders *et al.*, 1994a; Brdjanovic *et al.*, 1997), Filipe *et al.* (2001a) could not succeed to achieve full biological P-removal. As hypothesized by Satoh *et al.* (1994), the original microbial population in the activated sludge may also be an important factor to achieve a satisfactory PAO culture when operating lab-scale EBPR systems.

1.6. Mathematical modelling of the PAO-GAO competition

In the last decade, mathematical modelling of activated sludge systems has become an integral part of biological wastewater treatment (Henze *et al.*, 2000), often for optimization and prediction of process performance, and as a supporting tool for design (Henze *et al,*. 2000; Brdjanovic *et al.*, 2000; Meijer *et al.*, 2002; Gernaey *et al.*, 2004). In recent years, diverse modelling techniques have also been applied to get a better assessment about the microbial population dynamics that takes place in activated sludge systems (Oehmen *et al.*, 2007). Within the context of the PAO-GAO competition, several authors have developed mathematical models with the objective of getting a better understanding of the interaction between these two types of microorganisms (Manga *et al.*, 2001; Zeng *et al.*, 2003b; Yagci *et al.*, 2003, 2004; Whang *et al.*, 2007; Zhang *et al.*, 2008). Although these models are important tools that have helped to increase our knowledge, they have been built up based on a relatively narrow range of operational and/or environmental conditions, and the impact of one parameter on the PAO-GAO competition is evaluated independently of other factors. Thus, there is an increasing need to extend those models in order to be able to study and evaluate the combined effects of key factors that influence the PAO-GAO competition (such as the carbon source, pH and temperature). An integrated mathematical model, which takes into account the effect, dependence and influence of these factors on the metabolisms of PAO and GAO, may be an important, flexible and useful tool towards the optimization of the EBPR process through improving the understanding of the interaction among bacterial populations under different environmental and operational conditions.

1.7. Motivation and scope of the thesis

Because GAO compete for substrate with PAO, their presence and proliferation have been linked to the instability, suboptimal operation and deterioration of the EBPR process. The PAO-GAO competition has attracted the attention of scientists, engineers and treatment plant practitioners, and a considerable big effort has been paid to get a better understanding about the factors affecting the interaction between these two types of microorganisms. However, it is still not totally known how certain environmental and operating conditions affect their competition, influencing the stability and reliability of the EBPR systems. With a clear need to keep efficient and well-operated biological phosphorus removal processes in order to reduce the phosphorus loads discharged into surface water bodies (EEC, 2005), and recover and preserve our surrounding ecosystems, it becomes extremely important to find and suggest strategies and control measures to favour the development of PAO and suppress the growth of GAO in EBPR activated sludge systems.

Despite that temperature has been identified as an important environmental factor affecting the PAO-GAO competition, limited attention has been paid to the (short- and long-term) effects of temperature on the stoichiometry and kinetics of the anaerobic and aerobic metabolisms of GAO. Moreover, temperature has a strong influence on the biomass growth rate influencing the minimum required aerobic SRT. To execute short-term tests is important in order to evaluate the temperature effects on the metabolic rates without the influence of bacterial population changes that might occur with long-term studies. Meanwhile, long-term temperature effect tests should be carried out to account for potential population changes and adaptations that may occur from the continuous exposure of the microorganisms to certain temperature. Mostly based on lab-scale studies, the types and fractions of volatile fatty acids, temperature and pH have been observed to have an important effect on the PAO-GAO competition and, therefore, on the EBPR process reliability and stability. However, little is known about whether (and how) these factors affect the EBPR microbial populations at full-scale activated sludge systems. The current molecular techniques used for the identification of PAO and GAO, important for the operation of EBPR systems, require sophisticated and relatively expensive equipment, which limits a wider use, particularly in small wastewater treatment plants. In addition, most of the (lab-scale) research regarding the PAO-GAO competition has been carried out focusing on the single effect of one parameter without evaluating the combined effect of different (key) factors, what could be assessed potentially faster and more economically through mathematical modelling. Thus, the following questions arise:

   a. How does temperature affect the kinetics and stoichiometry of the anaerobic and aerobic metabolisms of GAO at short- and long-term?
   b. What are the factors affecting the PAO and GAO microbial populations in full-scale EBPR activated sludge systems?
   c. How could a practical method be developed for the quantification of PAO and GAO based on relatively simple analytical methods (such as orthophosphate and acetate)?
   d. How could mathematical modelling be used to study and evaluate the combined effect of key environmental and operating factors (such as temperature, the influent propionate to acetate ratios and pH levels) on the PAO-GAO competition?

1.8. Objectives

The main objective of the present PhD thesis is to evaluate the effect of key environmental and operating conditions (such as temperature, types and fractions of volatile fatty acids and pH levels) on the PAO-GAO competition, at lab- and full-scale systems and through mathematical modelling, in order to suggest potential strategies and control measures for the design and operation of more stable and reliable EBPR activated sludge systems.

The particular objectives are:

a. To determine the short- and long-term temperature effects and dependencies on the anaerobic and aerobic metabolisms of glycogen accumulating organisms, regarding their stoichiometry and kinetics.
b. To evaluate which environmental and operating conditions (factors) affect the PAO and GAO microbial populations at full-scale EBPR activated sludge systems in The Netherlands.
c. To develop a practical method, based on relatively common and reliable analytical techniques (such as acetate and orthophosphate), for the quantification of PAO and GAO populations at activated sludge wastewater treatment plants.
d. To study the combined effect of key environmental and operating conditions (like temperature, types and fractions of volatile fatty acids like acetate and propionate, and pH levels) on the PAO-GAO competition using mathematical modelling.

1.9. Outline of the thesis

The present dissertation comprises eight chapters. In the current chapter (chapter No. 1), a brief literature review has been presented concerning the PAO-GAO competition. In chapters 2 and 3, systematic studies regarding the effects of temperature on the anaerobic (chapter 2) and aerobic (chapter 3) metabolisms of an enriched GAO culture are provided. In those chapters, the temperature dependencies of the different involved metabolic processes are presented as well. The long-term temperature effects on the metabolism of an enriched GAO culture, required to evaluate potential population changes and adaptations, are discussed in chapter 4. Moreover, the temperature effects on the minimum aerobic SRT of GAO are also determined. Since most of the research concerning the PAO-GAO competition has focused on lab-scale studies, chapter 5 presents a full-scale survey undertaken to determine which factors could affect the PAO and GAO microbial populations at full-scale EBPR activated sludge systems in the Netherlands. In chapter 6, using highly enriched PAO and GAO cultures cultivated in two different lab-scale reactors, a practical method for the quantification of PAO and GAO is developed. In order to study and evaluate the combined effect of (key) environmental and operating conditions on the PAO-GAO competition, a mechanistic integrated metabolic model for PAO and GAO, which incorporates their temperature, carbon source (as acetate and propionate) and pH dependences, is presented in chapter 7. In the last chapter (chapter 8), overall conclusions, recommended strategies and control measures for the design and operation of EBPR activated sludge systems, as well as comments and ideas for further research, are discussed.

1.10. References

Amann RI (1995) *In situ* identification of microorganisms by whole cell hybridization with rRNA-targeted nucleic acid probes. In: Akkermans ADL, van Elsas JD, de Bruijn FJ, editors. Molecular microbial ecology manual. London: Kluwer Academic Publisher. p 1-15.

Barat R, van Loosdrecht MCM (2006) Potential phosphorus recovery in a WWTP with the BCFS ® process: interactions with the biological process. Wat Res 40(19): 3507-3516.

Barnard JL, Steichen MT. 2006. Where is nutrient removal going on? Water Sci Technol 53(3): 155-164.

Bazin MJ, Prosser JI (2000) Physiological models in microbiology. CRC Series in mathematical models in microbiology. Florida: CRC Press.

Beer M, Kong YH, Seviour RJ (2004) Are some putative glycogen accumulating organisms (GAO) in anaerobic: aerobic activated sludge systems members of the *α-Proteobacteria*? Microbiology 150: 2267–2275.

Brdjanovic D, van Loosdrecht MCM, Hooijmans CM, Alaerts GJ, Heijnen JJ. (1997) Temperature effects on physiology of biological phosphorus removal. ASCE J Environ Eng 123(2): 144-154.

Brdjanovic D, Logemann S, van Loosdrecht MCM, Hooijmans CM, Alaerts GJ, Heijnen JJ. (1998a) Influence of temperature on biological phosphorus removal: process and molecular ecological studies. Water Res 32(4): 1035-1048.

Brdjanovic D, Slamet A, van Loosdrecht MCM, Hooijmans CM, Alaerts GJ, Heijnen JJ (1998b) Impact of excessive aeration on biological phosphorus removal from wastewater. Wat Res 32(1): 200-208.

Brdjanovic D, van Loosdrecht MCM, Hooijmans CM, Alaerts GJ, Heijnen JJ. (1998c) Minimal aerobic sludge retention time in biological phosphorus removal systems. Biotech Bioeng 60(3): 326-332.

Brdjanovic D, van Loosdrecht MCM, Hooijmans C M, Mino T, Alaerts G J, Heijnen JJ (1999) Innovative methods for sludge characterization in biological phosphorus removal systems. Water Sci Technol 39(6): 37–43.

Brdjanovic, D., van Loosdrecht, M.C.M., Versteeg, P., Hooijmans, C.M., Alaerts, G.J., Heijnen, J.J. (2000) Modeling COD, N and P removal in a full-scale wwtp Haarlem Waarderpolder. Water Res 34(3): 846-858.

Bond PL, Erhart R, Wagner M, Keller J, Blackall LL (1999) Identification of some of the major groups of bacteria in efficient and nonefficient biological phosphorus removal activated sludge systems. Appl Environ Microbiol 65(9): 4077–4084.

Burow LC, Kong Y, Nielsen JL, Blackall LL, Nielsen PH (2007) Abundance and ecophysiology of *Defluviicoccus spp.* glycogen-accumulating organisms in full-scale wastewater treatment processes. Microbiology-SGM 153:178-185.

Cech JS, Hartman P (1993) Competition between phosphate and polysaccharide accumulating bacteria in enhanced biological phosphorus removal systems. Water Res 27: 1219-1225.

Chen Y, Randall AA, McCue T (2004) The efficiency of enhanced biological phosphorus removal from real wastewater affected by different ratios of acetic to propionic acid. Wat Res 38(1): 27-36.

Crocetti GR, Banfield JF, Keller J, Bond PL, Blackall LL (2002) Glycogen accumulating organisms in laboratory-scale and full-scale wastewater treatment processes. Microbiol 148: 3353-3364.

Crocetti GR, Hugenholtz P, Bond PL, Schuler A, Keller J, Jenkins D, Blackall LL (2000) Identification of polyphosphate-accumulating organisms and design of 16S rRNA-directed probes for their detection and quantitation. Appl Environ Microbiol 66(3): 1175–1182.

Dai Y, Yuan Z, Wang X, Oehmen A, Keller J (2007) Anaerobic metabolism of *Defluviicoccus vanus* related glycogen accumulating organisms (GAOs) with acetate and propionate as carbon sources. Wat Res 41(9): 1885-1896.

Daims H, Bruhl A, Amman R, Schleifer KH, Wagner M (1999) The domain-specific probe EUB 338 is insufficient for the detection of all bacteria: development and evaluation of a more comprehensive probe set. Syst Appl Microbiol 22: 434-444.

Dircks K, Beun JJ, van Loosdrecht MCM, Heijnen JJ, Henze M (2001) Glycogen metabolism in aerobic mixed cultures. Biotechnol Bioeng 73(2): 85-94.

EEA (2005) Source apportionment of nitrogen and phosphorus inputs into the aquatic environment. European Environmental Agency. Report No. 7. ISSN: 1725-9177. Copenhagen, Denmark.

Erdal UG, Erdal ZK, Randall CW (2003) The competition between PAO (phosphorus accumulating organisms) and GAO (glycogen accumulating organisms) in EBPR (enhanced biological phosphorus removal) systems at different temperatures and the effects on system performance. Water Sci Technol 47(11): 1-8.

EEC (1991) European Economic Community. Union Council Directive 91/271/EEC concerning urban waste water treatment. May 21st, 1991.

Filipe CDM, Daigger GT, Grady Jr CPL (2001a) A metabolic model for acetate uptake under anaerobic conditions by glycogen-accumulating organisms: stoichiometry, kinetics and effect of pH. Biotechnol Bioeng 76(1): 17-31.

Filipe CDM, Daigger GT, Grady Jr CPL (2001b) Effects of pH on the aerobic metabolism of phosphate-accumulating organisms and glycogen-accumulating organisms. Water Environ Res 73(2): 213-222.

Filipe CDM, Daigger GT, Grady Jr CPL (2001c) pH as a key factor in the competition between glycogen-accumulating organisms and phosphorus-accumulating organisms. Water Environ Res 73(2): 223-232.

Filipe CDM, Daigger GT, Grady Jr CPL (2001d) Stoichiometry and kinetics of acetate uptake under anaerobic conditions by an enriched culture of phosphorus-accumulating organisms at different pH. Biotechnol Bioeng 76(1): 32-43.

Gernaey KV, van Loosdrecht MCM, Henze M, Lind M, Jorgensen SB (2004) Activated sludge wastewater treatment plant modellign and simulation: state of the art. Environ Modelling Software 19(9):763-783.

Gijzen H J, Mulder A (2001). The nutrient cycle of balance. Water 21, August 2001, 38-40.

Grady Jr. CPL, Filipe CDM. 2000. Ecological engineering from biorreactors for wastewater treatment. Water Air Soil Pollut 123: 117-132.

Gu, A.Z., Saunders, A.M., Neethling, J.B., Stensel, H.D., Blackall, L. (2005) In: WEF (Ed.) Investigation of PAOs and GAOs and their effects on EBPR performance at full-scale wastewater treatment plants in US, October 29–November 2, WEFTEC, Washington, DC, USA.

Henze M, Gujer W, Mino T, van Loosdrecht MCM. (2000) Activated sludge models ASM1, ASM2, ASM2d and ASM3. IWA Scientific and Technical Report No. 9. IWA Task Group on Mathematical Modelling for Design and Operation of Biological Wastewater Treatment. London: IWA Publishing.

Hesselmann RPX, Werlen C, Hahn D, van der Meer JR, Zehnder AJB (1999) Enrichment, phylogenetic analysis and detection of a bacterium that performs enhanced biological phosphate removal in activated sludge. Syst Appl Microbiol 22(3): 454–465.

Jansen PMJ, Meinema K, van der Roest HF (2002) Biological phosphorus removal. Manual for design and operation. Water and wastewater practitioner series: STOWA report. Ed. IWA publishing. ISBN: 1843390124. Cornwall, UK.

Jobbàgy A, Litherathy B, Tardy G. 2002. Implementation of glycogen accumulating bacteria in treating nutrient-deficient wastewater. Water Sci Technol 46(1-2):185-190.

Kong YH, Ong SL, Ng WJ, Liu WT (2002) Diversity and distribution of a deeply branched novel proteobacterial group found in anaerobic–aerobic activated sludge processes. Environ Microbiol 4(11): 753–757.

Kong YH, Nielsen JL, Nielsen PH (2005) Identity and ecophysiology of uncultured actinobacterial polyphosphate-accumulating organisms in full-scale enhanced biological phosphorus removal plants. Appl. Environ. Microbiol. 71(7): 4076–4085.

Korstee GJJ, Appeldoorn KJ, Bonting CFC, van Niel EWJ, van Veen HW (1994) Biology of polyphosphate-accumulating bacteria involved in enhanced biological phosphorus removal. FEMS Microbiology Revs. 15: 137-153.

Kuba T, Murnleiter E, van Loosdrecht MCM, Heijnen JJ (1996) A metabolic model for biological phosphorus removal by denitrifying organisms. Biotechnol Bioeng 52(6): 685-695.

Kuba T, Wachmeister A, van Loosdrecht MCM, Heijnen JJ (1994) Effect of nitrate on phosphorus release in biological phosphorus removal systems. Wat Sci Technol 30(6): 263-269.

Liu WT, Mino T, Nakamura K, Matsuo T (1994) Role of glycogen in acetate uptake and polyhydroxyalkanoate synthesis in anaerobic-aerobic activated sludge with a minimized polyphosphate content. J Ferment Bioeng 77(5): 535-540.

Liu WT, Mino T, Nakamura K, Matsuo T (1996) Glycogen accumulating population and its anaerobic substrate uptake in anaerobic-aerobic activated sludge without biological phosphorus removal. Water Res 30(1): 75-82.

Liu WT, Nakamura K, Matsuo T, Mino T. (1997) Internal energy-based competition between polyphosphate- and glycogen-accumulating bacteria in biological phosphorus removal reactors-effect of P/C feeding ratio. Water Res 31(6): 1430-1438.

Lopez C, Pons MN, Morgenroth E (2006) Endogenous processes during long-term starvation in activated sludge performing enhanced biological phosphorus removal. Wat Res 40: 1519 – 1530.

Lu H, Oehmen A, Virdis B, Keller J, Yuan Z (2006) Obtaining highly enriched cultures of *Candidatus Accumulibacter Phosphatis* through alternating carbon sources. Water Res 40(20): 3838-3848.

Mamais D, Jenkins D (1992) The effects of MCRT and temperature on enhanced biological phosphorus removal. Wat Sci Tech 26(5-6):955-965.

Manga J, Ferrer J, Garcia-Usach F, Seco A (2001) A modification to the Activated Sludge Model No. 2. Water Sci Technol 43(11): 161-171.

Matsuo Y, Kitagawa M, Tanaka T, Miya A (1982) Sewage and night-soil treatment by anaerobic aerobic activated sludge processes. Proceedings Environ Sanitary Eng Res 19:82-87 (in Japanese).

Matsuo Y (1994) Effect of the anaerobic solids retention time on enhanced biological phosphorus removal. Water Sci Technol 30(6): 193-202.

Meijer SCF, van Loosdrecht MCM, Heijnen JJ (2001) Metabolic modelling of full-scale biological nitrogen and phosphorus removing WWTP's. Water Res 35(11): 2711-2723.

Meijer SCF, van Loosdrecht MCM, Heijnen JJ (2002) Modelling the start-up of a full-scale biological nitrogen and phosphorus removing WWTP's. Water Res 36(11): 4667-4682.

Metcalf and Eddy (2003) Wastewater engineering. Treatment and reuse. 4rd. ed. Ed. Mc Graw Hill.

Meyer RL, Saunders AM, Blackall LL (2006) Putative glycogen accumulating organisms belonging to *Alphaproteobacteria* identified through rRNA-based stable isotope probing. Microbiology-SGM 152: 419–429.

Mino T, Liu WT, Kurisu F, Matsuo T (1995) Modeling glycogen storage and denitrification capability of microorganisms in enhanced biological phosphate removal processes. Water Sci Technol 31(2): 25–34.

Mino T, van Loosdrecht MCM, Heijnen JJ (1998) Microbiology and biochemistry of the enhanced biological phosphorus removal process. Water Res 32(11): 3193-3207.

Murnleitner E, Kuba T, van Loosdrecht MCM, Heijnen JJ (1997) An integrated metabolic model for the aerobic and denitrifying biological phosphorus removal. Biotechnol Bioeng 54(5): 434-450.

Nielsen AT, Liu WT, Filipe C, Grady L, Molin S, Stahl DA (1999) Identification of a novel group of bacteria in sludge from a deteriorated biological phosphorus removal reactor. Appl Environ Microbiol 65(3): 1251–1258.

Oehmen A (2004) The competition between polyphosphate accumulating organisms and glycogen accumulating organisms in the enhanced biological phosphorus removal process. PhD Thesis. The University of Queensland. Brisbane, Australia.

Oehmen A, Vives MT, Lu H, Yuan Z, Keller J (2005a) The effect of pH on the competition between polyphosphate-accumulating organisms and glycogen-accumulating organisms. Water Res 39(15): 3727-3737.

Oehmen A, Yuan Z, Blackall LL, Keller J (2005b) Comparison of acetate and propionate uptake by polyphosphate accumulating organisms and glycogen accumulating organisms. Biotechnol Bioeng 91(2): 162-168.

Oehmen A, Zeng RJ, Yuan Z, Keller J (2005c) Anaerobic metabolism of propionate by polyphosphate-accumulating organisms in enhanced biological phosphorus removal systems. Biotechnol Bioeng 91(1): 43-53.

Oehmen A, Saunders AM, Vives MT, Yuan Z, Keller J (2006a) Competition between polyphosphate and glycogen accumulating organisms in enhanced biological phosphorus removal systems with acetate and propionate as carbon sources. J Biotechnol 123(1): 22-32.

Oehmen A, Zeng RJ, Saunders AM, Blackall LL, Keller J, Yuan Z (2006b) Anaerobic and aerobic metabolism of glycogen accumulating organisms selected with propionate as the sole carbon source. Microbiology 152(9): 2767-2778.

Oehmen A, Lemos PC, Carvalho G, Yuan Z, Keller J, Blackall LL, Reis MAM (2007a) Advances in enhanced biological phosphorus removal: from micro to macro scale. Water Res 41(11): 2271-2300.

Oehmen A, Zeng RJ, Keller J, Yuan Z (2007b) Modeling the aerobic metabolism of polyphosphate-accumulating organisms enriched with propionate as a carbon source. Water Environ Res 79(13): 2477-2486.

Olsson G (2006) Instrumentation, control and automation in the water industry – state-of-the-art and new challenges. Wat Sci Tech 53(4-5): 1-16.

Panswad T, Doungchai A, Anotai J (2003) Temperature effect on microbial community of enhanced biological phosphorus removal system. Water Res 37: 409-415.

Park JK, Ahn CH, Whang LM, Lee EK, Lee YO, Probst T, Reichardt RN, Brauer J W. 2001. Problems in biological phosphorus removal with dairy wastewater. Proceedings of the 7th Annual Industrial Wastes Technical and Regulatory Conference. Water Environment Federation. August 12–15, 2001, Charleston.

Pijuan M, Saunders AM, Guisasola A, Baeza JA, Casas C, Blackall LL (2004) Enhanced biological phosphorus removal in a sequencing batch reactor using propionate as the sole carbon source. Biotechnol Bioeng 85(1): 56-67.

Puig S, Corominas LI, Balaguer MD, Colprim J (2007) Biological nutrient removal by applying SBR technology in small wastewater treatment plants: carbon source and C/N/P ratio effects. Wat Sci Technol 55(7): 135-141.

Reichert P (1994) AQUASIM - a tool for simulation and data analysis of aquatic systems. Water Sci Technol 30(2): 21-30.

Rodrigo MA, Seco A, Ferrer J, Penya-Roja JM (1999) The effect of the sludge age on the deterioration of the enhanced biological phosphorus removal process. Environ Tech 20(10):1055-1063.

Satoh H, Mino T, Matsuo T (1994) Deterioration of enhanced biological phosphorus removal by the domination of microorganisms without polyphosphate accumulation. Water Sci Technol 30(6): 203-211.

Saito T, Brdjanovic D, van Loosdrecht MCM (2004) Effect of nitrite on phosphate uptake by phosphate accumulating organisms. Water Res 38(17): 3760–3768.

Saunders AM, Oehmen A, Blackall LL, Yuan Z, Keller J (2003) The effect of GAO (glycogen accumulating organisms) on anaerobic carbon requirements in full-scale Australian EBPR (enhanced biological phosphorus removal) plants. Water Sci Technol 47(11): 37-43.

Schuler AJ, Jenkins D (2002) Effects of pH on enhanced biological phosphorus removal metabolisms. Wat Sci Technol 46(4-5): 171-178.

Schuler AJ, Jenkins D (2003) Enhanced biological phosphorus removal from wastewater by biomass with different phosphorus contents, Part 1: Experimental results and comparison with metabolic models. Water Environ Res 75(6): 485-498.

Seviour RJ, Blackall LL (1999) The microbiology of activated sludge. Kluwer. Dordrecht, The Netherlands.

Seviour RJ, Mino T, Onuki M (2003) The microbiology of biological phosphorus removal in activated sludge systems. FEMS Microbiol Rev 27(1): 99-27.

Smolders GJF, van der Meij J, van Loosdrecht MCM, Heijnen JJ (1994a) Model of the anaerobic metabolism of the biological phosphorus removal process: stoichiometry and pH influence. Biotechnol Bioeng 43(6): 461-470.

Smolders GJF, van der Meij J, van Loosdrecht MCM, Heijnen JJ (1994b) Stoichiometric model of the aerobic metabolism of the biological phosphorus removal process. Biotechnol Bioeng 44: 837-848.

Smolders GJF, van der Meij J, van Loosdrecht MCM, Heijnen JJ (1995) A structured metabolic model for anaerobic and aerobic stoichiometry and kinetics of the biological phosphorus removal process. Biotechnol Bioeng 47(3): 277-287.

Smolders GJF (1995) A metabolic model of the biological phosphorus removal: stoichiometry, kinetics and dynamic behaviour. PhD thesis. Delft University of Technology. Delft, The Netherlands.

Thomas M, Wright P, Blackall L, Urbain V, Keller J (2003) Optimisation of Noosa BNR plant to improve performance and reduce operating costs. Water Sci Technol 47 (12): 141–148.

van Loosdrecht MCM, Hooijmans CM, Brdjanovic D, Heijnen JJ (1997) Biological phosphorus removal processes: a mini review. Applied microbiology and biotechnology 48: 289-296.

van Veldhuizen HM, van Loosdrecht MCM, Heijnen JJ (1999) Modelling biological phosphorus and nitrogen removal in a full scale activated sludge process. Water Res 33(16): 3459-3468.

van Niel EWJ, Appeldoorn KJ, Zehnder AJB, Kortstee GJJ (1998) Inhibition of anaerobic phosphate release by nitric oxide in activated sludge. App Environ Microbiol 64(8): 2925-2930.

Whang LM, Filipe CDM, Park JK (2007) Model-based evaluation of competition between polyphosphate- and glycogen-accumulating organisms. Water Res 41(6): 1312-1324.

Whang LM, Park JK (2006) Competition between polyphosphate- and glycogen-accumulating organisms in enhanced biological phosphorus removal systems: effect of temperature and sludge age. Water Environ Res 78(1): 4-11.

Wilderer PA, Irvine RL, Goronszy MC (2001). Sequencing batch reactor technology. Scientific and technical report No. 10. IWA Publishing.

Wong, MT, Tan FM, Ng WJ, Liu WT (2004) Identification and occurrence of tetrad-forming *Alphaproteobacteria* in anaerobic–aerobic activated sludge processes. Microbiology-SGM 150: 3741–3748.

Wong MT, Mino T, Seviour RJ, Onuki M, Liu WT (2005) *In situ* identification and characterization of the microbial community structure of full-scale enhanced biological phosphorous removal plants in Japan. Water Res 39(13): 2901–2914.

Yagci N, Artan N, Cogkor EU, Randall C, Orhon D (2003) Metabolic model for acetate uptake by a mixed culture of phosphate- and glycogen-accumulating organisms under anaerobic conditions. Biotechnol Bioeng 84(3): 359-373.

Yagci N, Insel G, Orhon D (2004) Modelling and calibration of phosphate and glycogen accumulating organism competition for acetate uptake in a sequencing batch reactor. Water Sci Technol 50(6): 241-250.

Ydstebϕ L, Bilstad T, Barnard J (2000) Experience with biological nutrient removal at low temperatures. Water Environ Res 72(4): 444-454.

Zeng RJ, van Loosdrecht MCM, Yuan Z, Keller J (2002) Proposed modifications to metabolic model for glycogen accumulating organisms under anaerobic conditions. Biotechnol Bioeng 80(3): 277-279.

Zeng RJ, van Loosdrecht MCM, Yuan Z, Keller J. (2003a) Metabolic model for glycogen-accumulating organisms in anaerobic/aerobic activated sludge systems. Biotechnol Bioeng 81(1): 92-105.

Zeng RJ, Yuan Z, Keller J (2003b) Model-based analysis of anaerobic acetate uptake by a mixed culture of polyphosphate-accumulating and glycogen-accumulating organisms. Biotechnol Bioeng 83(3): 293-302.

Zhang C, Chen Y, Randall AA, Gu G. (2008) Anaerobic metabolic models for phosphorus- and glycogen- accumulating organisms with mixed acetic and propionic acids as carbon sources, Water Res (in press), doi: 10.1016/j.waters.2008.06.025.

Zilles JL, Peccia J, Kim MW, Hung CH, Noguera DR (2002) Involvement of *Rhodocyclus*-related organisms in phosphorus removal in full-scale wastewater treatment plants. Appl Environ Microbiol 68(6): 2763–2769.

2
Chapter

Temperature effects on
the anaerobic metabolism of GAO

Content

This chapter has been published as:

    Lopez-Vazquez CM, Song YI, Hooijmans CM, Brdjanovic D, Moussa MS, Gijzen HJ, van Loosdrecht MCM (2007) Short-term temperature effects on the anaerobic metabolism of Glycogen Accumulating Organisms. *Biotechnol Bioeng*, 97(3):483-495.

Abstract

The effects of temperature on the anaerobic metabolism of GAO were studied in a broad temperature range (from 10 to 40 °C). Additionally, maximum acetate uptake rate of PAO, between 20 and 40 °C, were also evaluated. It was found that GAO had clear advantages over PAO for substrate uptake at temperature higher than 20 °C. Below 20 °C, maximum acetate uptake rates of both microorganisms were similar. However, lower maintenance requirements at temperature lower than 30 °C give PAO metabolic advantages in the PAO-GAO competition. Consequently, Polyphosphate-accumulating organisms could be considered to be psychrophilic microorganisms while glycogen-accumulating organisms appear to be mesophilic. These findings contribute to understand the observed stability of the EBPR process in wastewater treatment plants operated under cold weather conditions. They may also explain the proliferation of GAO in wastewater treatment plants and thus, EBPR instability, observed in hot climate regions or when treating warm industrial effluents. It is suggested to take into account the observed temperature dependencies of PAO and GAO in order to extend the applicability of current activated sludge models to a wider temperature range.

2.1. Introduction

Most of the reports that describe the EBPR process deterioration and the dominance of GAO at full-scale wastewater treatment plants (Grady and Filipe, 2000; Crocetti *et al.*, 2002; Saunders *et al.*, 2003; Thomas *et al.*, 2003; Wong *et al.*, 2005; Barnard and Steichen, 2006) are from regions where usually the average yearly atmospheric temperature is around or higher than 25 °C. Therefore, higher wastewater temperature could be expected (> 20 °C). Besides, the dominance of GAO has also been reported in wastewater treatment plants handling warm industrial effluents (where wastewater temperature is higher than 27 °C) (Park *et al.*, 2001; Jobàggy *et al.*, 2002). At lab-scale, different researchers have studied the temperature effects on the biological phosphorus removal (BPR) process to explain the effects of temperature on the PAO-GAO competition (Panswad *et al.*, 2003; Erdal *et al.*, 2003; Whang and Park, 2002, 2006). In general, these reports agree on the statement that, at higher temperature (> 20 °C), the activity of the BPR process tends to deteriorate and GAO become the dominant microorganisms. However, since these studies were not performed with enriched PAO and GAO cultures, the underlying mechanisms of the EBPR process deterioration and actual temperature effects on the metabolism of PAO and GAO remain unclear. It is still unknown whether the EBPR deterioration occurred because: a) PAO are not able to adapt to higher temperature levels (>20 °C) giving GAO the chance to proliferate, b) GAO have metabolic advantages over PAO at high temperatures, or c) a combination of both previous causes.

Brdjanovic *et al.* (1997, 1998a) carried out a systematic study on an enriched PAO culture in order to understand the short- and long-term temperature effects on the EBPR process from 5 to 30 °C. According to their observations, the observed temperature dependencies at short- (a few hours) and long-term (weeks) were well described by applying the Arrhenius temperature coefficients calculated in that research. The Arrhenius coefficients obtained at short- and long-term were similar but not applicable for the whole temperature interval of that study (Brdjanovic *et al.*, 1998a). Although the overall aerobic Arrhenius temperature coefficient was valid for their whole studied temperature range (from 5 to 30 °C), the overall anaerobic temperature coefficient was found to be only valid up to 20 °C. An analogous study with an enriched GAO culture has not been reported yet. Since substrate uptake and, therefore the PAO-GAO competition, occurs under anaerobic conditions then the effects of temperature on their anaerobic metabolisms play a crucial role.

The aims of this study were: 1) to evaluate the short-term (hours) temperature effects on the anaerobic metabolism, regarding stoichiometry and kinetics, of an enriched GAO culture within a relatively broad temperature range (from 10 to 40 °C); and, 2) to partially repeat (from 20 to 30 °C) and extend (up to 40 °C) the short-term tests carried out by Brdjanovic *et al.* (1997) on the anaerobic acetate uptake rate of PAO. Consequently, the PAO-GAO competition was also studied based on the temperature effects on their anaerobic acetate uptake rates and anaerobic maintenance requirements. These could provide important information to understand the behaviour of PAO and GAO, the mechanisms of their competition and, consequently, the EBPR stability, within an ample temperature interval (from 10 to 40 °C) that covers the operating temperature range of most of the domestic and industrial wastewater treatment plants.

2.2. Materials and methods

2.2.1. Continuous operation of the sequencing batch reactors

GAO and PAO cultures were enriched in two separate double-jacketed lab-scale sequencing batch reactors (SBRs). Each SBR had a working volume of 2.5 L. Activated sludge from domestic wastewater treatment plants (Hoek van Holland and Haarlem Waarderpolder, The Netherlands) was used as inoculum. SBRs were operated at 20 ± 0.5 °C in cycles of 6 hours (2.25 h anaerobic, 2.25 h aerobic and 1.5 h settling phase) following the same operating conditions applied in previous reports (Smolders *et al.*, 1994a; Brdjanovic *et al.*, 1997). pH was maintained at 7.0 ± 0.1 by dosing 0.3 M HCl and 0.2 M NaOH. In order to create the anaerobic conditions, nitrogen gas was supplied in the anaerobic phase at a flow rate of 30 L/h while air was provided during the aerobic stage at a flow rate of 60 L/h. The cycle started with the supply of nitrogen gas for 5 minutes in order to remove oxygen remaining from the previous cycle and feed the medium under no oxygen presence. Therefore, after the first 5 minutes of the cycle, 1.25 L of synthetic medium was fed to the SBRs over a period of 5 minutes. 250 mL of mixed liquour were removed on a daily basis (62.5 mL per cycle) from the SBRs, resulting in a sludge retention time (SRT) of approximately 10 d. The sludge waste was weekly corrected due to biomass loss through the effluent. At the end of the settling period 1.19 L of supernatant was pumped out from the reactors, resulting in a hydraulic retention time (HRT) of 12 h. The reactors were constantly mixed at 500 rpm except during settling and decant phases. Anaerobic batch experiments at different temperatures were performed in a separate reactor after the biomass activity reached steady-state conditions. Thus, the two SBRs were used only as sources of enriched GAO and PAO biomass.

2.2.2. Operation of the batch reactor

A double-jacketed laboratory fermenter with a maximal volume of 0.5 L was used for the execution of the anaerobic batch experiments. The experiments were performed at controlled temperature and pH (7.00 ± 0.05). pH was maintained by dosing 0.2 M HCl and 0.2 M NaOH. At the beginning of each experiment, enriched GAO or PAO sludge (according to the corresponding experiment) was manually transferred from the respective SBR to the batch reactor. The amount of transferred sludge was 250 mL, which is the amount of sludge daily wasted from each SBR, to avoid disturbing the continuous operation and steady-state conditions of the two main SBRs. The sludge used in the batch tests was not returned to the SBRs. N_2 gas was continuously introduced to the reactor during the whole test at a flow rate of 30 L/h. During the anaerobic batch experiments, the sludge in the batch reactor was constantly stirred at 500 rpm.

2.2.3. Synthetic media

The main difference between the operating conditions of the GAO and PAO SBRs was the phosphorus content in the synthetic medium supplied to each reactor. Phosphorus concentration in GAO SBR influent was limited to 2.2 mg PO_4^{3-}-P/L (0.07 P-mmol/L) (Liu *et al.*, 1997). While synthetic medium used for PAO SBR contained 15 mg PO_4^{3-}-P/L (0.48 P-mmol/L) (Smolders *et al.*, 1994a). Besides the different phosphorus concentration, synthetic media contained per litre: 850 mg NaAc·3H$_2$O (12.5 C-mmol, approximately 400 mg COD/L) and 107 mg NH$_4$Cl (2 N-mmol). 0.002 g of allyl-N thiourea (ATU) were added to inhibit nitrification. This concentration has shown to be sufficient to inhibit the growth of nitrifying organisms in lab-EBPR systems (Zeng *et al.*, 2003) likely because the SBR operating conditions do not favor nitrification and due to the continuous ATU addition. The rest of minerals and trace metals present in the synthetic media were prepared as described by Smolders *et al.*, (1994a). Prior to use, synthetic media were autoclaved at 110 °C for 1 h.

2.2.4. Analyses

The performance of the GAO and PAO SBRs was regularly monitored by measuring orthophosphate (PO_4^{3-}-P), mixed liquor suspended solids (MLSS) and mixed liquor volatile suspended solids (MLVSS). In batch experiments, orthophosphate (PO_4^{3-}P), acetate (HAc), MLSS, MLVSS, poly-hydroxy-butyrate (PHB), poly-hydroxy-valerate (PHV), glycogen (Gly) and ammonium (NH^{4+}-N) concentrations were measured. Orthophosphate (PO_4^{3-}-P) was determined by the ascorbic acid method, ammonia and nitrate were measured spectrophotometrically. All analyses, including the MLSS and MLVSS determination, were performed in accordance with Standard Methods (APHA, 1998). The PHA content (as PHB and PHV) of the freeze dried biomass was determined according to the method described by Smolders *et al.* (1994a). Glycogen was also determined following the method described by Smolders *et al.* (1994a) but extending the digestion phase at 100°C from 1 to 5 h.

Fluorescence *in situ* Hybridization (FISH) was performed as described in Amman (1995). The EUBMIX probe (mixture of probes EUB 338, EUB338-II and EUB338-III) to target the entire bacterial population, PAOMIX probe (mixture of probes PAO462, PAO651 and PAO 846) to target *Accumulibacter* (Crocetti *et al.*, 2000) and GAOMIX probe (mixture of probes GAOQ431 and GAOQ989) to target *Competibacter* (Crocetti *et al.*, 2002) were used to determine the microbial population distribution of the GAO culture. Hybridization conditions of the FISH samples were performed according to Crocetti *et al.* (2000, 2002).

The quantification of the population distribution was carried out using the MATLAB image processing toolbox (The Mathworks, Natick, MA). 8-bit images for each of the color channels (red for GAO, green for PAO and blue for EUB) were converted into binary format using direct thresholding at a graylevel determined using the Otsu method (Otsu, 1979), where pixels with value below the threshold level represent the background. Image coverage was computed by dividing the number of pixels corresponding to the object with the total number of pixels of the image.

2.2.5. Anaerobic stoichiometry and acetate uptake rate of GAO

As previously described, at the end of the aerobic phase 250 mL of the GAO culture enriched at 20 °C were manually transferred to the batch reactor. The working temperature of the batch reactor (10, 15, 20, 25, 30, 35 and 40 °C) was set 1 h before the sludge transfer as well as the temperature of the synthetic medium. In order to adjust the microorganisms to the new temperature, the sludge was stirred and aerated for 1 h before starting the anaerobic batch test. N_2 gas was supplied during the whole experiment to keep anaerobic conditions. After exposing the sludge to anaerobic conditions for 5 minutes, synthetic medium was added to the reactor in a pulse mode. The amount of acetate in the synthetic medium previously described was adjusted according to the set temperature in order to obtain a full HAc-uptake without changing the 2 h length and considering the biomass concentration dilution due to the addition of medium. Therefore, the initial HAc concentration in the first batch tests (20 and 25 °C) was reduced from the initial HAc concentration in the SBR of 6.25 C-mmol/L to 3.10 C-mmol HAc/L. However, as the studied temperature increased, GAO exhibited a higher HAc uptake rate. Consequently, at 30, 35 and 40 °C the initial HAc concentration was increased to 6.25 (30 °C) and 9.38 C-mmol/L (35 and 40 °C) to allow sufficient time for an accurate determination of the maximum HAc uptake rate. An extensive characterization of both liquid phase and biomass was performed during the anaerobic batch tests.

Glycogen hydrolysis and PHB and PHV production per HAc consumed (Gly/HAc, PHB/HAc, PHV/HAc ratios, respectively) were the stoichiometric parameters of interest in this study. These parameters were calculated taking into account the initial HAc concentration and biomass composition (as MLSS, MLVSS, Gly, PHB and PHV) at the beginning and end of each batch experiment.

The maximum acetate consumption rate $q_{SA,GAO}^{MAX}$ for the different studied temperatures, was determined considering the HAc consumption profile and active biomass concentration at the beginning of the corresponding batch experiment. Specific acetate uptake rates were expressed in C-mol/C-mol Biomass/h units. The active biomass was determined as total volatile suspended solids concentration but excluding PHB, PHV and glycogen contents (active biomass = MLVSS – PHB – PHV – Glycogen). Active biomass was used in order to distinguish between biomass and organic and inorganic storage products since biomass (either PAO or GAO) can be made up for 50 % of internal storage products. In addition, fractions of PHB, PHV and Glycogen change strongly due to dynamics in the anaerobic and aerobic phase.

In order to calculate the specific rates, active biomass concentration was expressed in C-mol units by taking into account the GAO biomass composition ($CH_{1.84}O_{0.5}N_{0.19}$) experimentally determined by Zeng *et al.* (2003).

2.2.6. Anaerobic ATP maintenance coefficient of GAO

At the end of the aerobic phase, 250 mL of enriched culture at 20 °C was manually transferred from the GAO SBR to the batch reactor. A working temperature in the batch reactor (10, 20, 25, 30, 35 and 40 °C) was set 1 h before the sludge transfer. Sludge was exposed to anaerobic conditions without substrate addition for 7 hours and PHB, PHV and glycogen were measured every 30 minutes.

The energy required to covering the anaerobic maintenance needs, as adenosine triphosphate (ATP) per active biomass per hour, is expressed in terms of the specific anaerobic maintenance coefficient $m^{an}_{ATP,GAO}$ (in mol ATP/C-mol biomass/h units). In previous reports (Filipe *et al.*, 2001a; Zeng *et al.*, 2003), it was proposed that glycogen is hydrolyzed to provide the required energy for anaerobic maintenance and is directly converted to PHB and PHV (Equation 2.1). Therefore, $m^{an}_{ATP,GAO}$ was determined, for each temperature, considering the anaerobic glycogen hydrolysis rate of each batch experiment and assuming that 1 C-mol of glycogen hydrolyzed produces 0.5 mol ATP (Zeng *et al.*, 2003). Corresponding glycogen hydrolysis rates were calculated based on the slope of the glycogen depletion profile observed in each batch test carried out without substrate addition. The active GAO biomass concentration, calculated as previously described, was used to calculate the specific rates. Hydrolysis rates were confirmed based on the PHB and PHV production profiles.

$$1\, Cmol\, Glycogen \longrightarrow \frac{1}{6}PHB + \frac{5}{12}PHV + \frac{1}{4}PH_2MV + \frac{1}{6}CO_2 \qquad (2.1)$$

2.2.7. Anaerobic acetate uptake rate of PAO

Four independent anaerobic batch experiments were performed to determine the short-term temperature effects on the anaerobic acetate uptake rate of PAO $q^{MAX}_{SA,PAO}$ (in C-mol/C-mol biomass/h units) at 20, 25, 30 and 35 °C. The anaerobic batch experiments with the PAO culture, enriched at 20 °C, were carried out following the same procedure previously described for the GAO biomass. The initial HAc concentration for the PAO anaerobic batch tests was kept constant at 6.25 C-mmol/L. Active biomass concentration was used for the determination of the specific rates and expressed in C-mol units by considering the PAO biomass composition ($CH_{2.09}O_{0.54}N_{0.20}P_{0.015}$) determined by Smolders *et al.* (1994b).

2.2.8. Temperature coefficients

The effect of temperature on a constant rate relative to a standard temperature (here 20 °C) can be expressed by the simplified Arrhenius equation:

$$r_T = r_{20} \cdot \theta_1^{(T-20)} \tag{2.2}$$

Where the r_T is the reaction at the temperature T, T is the temperature in °C and θ_1 is the temperature coefficient.

In this study the simplified Arrhenius equation is used. It allows comparing results with the temperature coefficients of different processes considered in mathematical models (Henze *et al.*, 2000). Furthermore, the equation is also well suited for fitting the results. In order to describe the kinetics of GAO for the whole experimental temperature range (from 10 to 40 °C), the simplified Arrhenius equation (Equation 2.2) was extended by including an additional mathematical term, which includes an extra Arrhenius coefficient (θ_2) to describe the declination in activity at superoptimal temperatures (modified from Bazin and Prosser, 2000). The extended Arrhenius equation is:

$$r_T = r_{20} \cdot \theta_1^{(T-20)} \left[1 - \theta_2^{(T-T_{MAX})} \right] \tag{2.3}$$

Where the r_T is the reaction at the temperature T, T is the temperature in °C, θ_1 is the temperature coefficient θ calculated from Equation 2.2, T_{MAX} is the temperature at which the microbial activity ceases, and θ_2 is a second temperature coefficient used to describe the declination in activity of the microorganisms.

2.3. Results

2.3.1. Performance of GAO SBR

The GAO SBR was continuously operated for more than 100 days before the measurements started. The biomass activity reached steady-state conditions within the first 50 days of operation. MLSS and MLVSS concentrations, at the end of the aerobic phase, were 3190 and 2879 mg/L, respectively, resulting in an average MLVSS/MLSS ratio of 0.90. High MLVSS/MLSS ratio indicated low inorganic matter storage, such as poly-P, which suggested that GAO were the dominant microrganisms. Cycle measurements carried out to determine the biomass activity showed the typical carbon profiles found in GAO enriched cultures (Figure 2.1): HAc uptake under anaerobic conditions associated by the consumption of glycogen, PHA production (as PHB and PHV) and a practically negligible anaerobic P/HAc ratio (0.01 P-mol/HAc C-mol) compared to typical ratios of enriched PAO cultures (0.50 P-mol/HAc C-mol, Smolders *et al.*, 1994a). Under aerobic conditions, the PHA previously stored in the anaerobic phase was oxidized and glycogen was produced. FISH analysis confirmed that GAO were the predominant microorganisms (Figure 2.2). A quantification of the microbial population distribution shown in Figure 2.2 indicated that, approximately, GAO were 75 % and PAO around 20 % of the total microbial population present in the GAO SBR and other bacteria 5 %. Based on the low ash content (through a high MLVSS/MLSS ratio), biomass activity, and confirmed with the FISH technique, it can be concluded that GAO were the dominant microorganisms in GAO SBR.

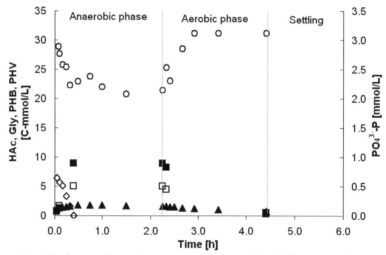

Figure 2.1. Cycle profiles observed in the enriched Glycogen Accumulating Organisms culture at 20 °C: acetate (◊), glycogen (○), PHB (■), PHV (□) and phosphate (▲).

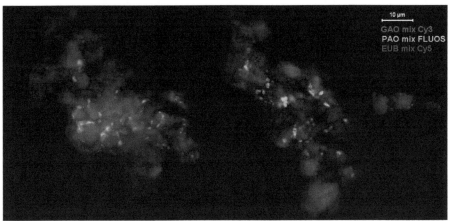

Figure 2.2. Bacterial population distribution in the enriched culture of Glycogen Accumulating Organisms by applying Fluorescence *in situ* Hybridization techniques (bar indicates 10 μm; blue: Eubacteria, red: Glycogen Accumulating Organisms and green: Polyphosphate Accumulating Organisms).

2.3.2. Anaerobic stoichiometry of GAO

The anaerobic conversions of relevant compounds were evaluated for each studied temperature. In particular, the anaerobic parameters determined at 20 °C were compared with other reports carried out with enriched GAO and PAO cultures since all of them were performed at the same temperature. Table 2.1 shows that most of the observed parameter values of this study (glycogen hydrolysis and PHA formation per HAc consumed as well as the PHV/PHB ratio) are in the range of previously reported ratios for GAO cultures and differ from reported anaerobic ratios for enriched PAO cultures. The glycogen/HAc uptake (Gly/HAc) ratio of this study (-1.20 ± 0.19 C-mol/C-mol) is similar to the theoretical ratio (-1.12 C-mol/C-mol) of the GAO anaerobic model proposed by Zeng *et al.* (2003). This theoretical anaerobic metabolic model was built up based on known biochemical pathways and experimentally validated with an enriched GAO culture. The PHA (PHA/HAc), PHB (PHB/HAc) and PHV (PHV/HAc) production ratios also follow similar trends when they are compared with the same reports. These results confirm that GAO were the dominant microorganisms in the enriched culture.

Table 2.1. Comparison between the anaerobic stoichiometric parameters observed in the present study on the culture of Glycogen Accumulating Organisms and other reports at 20 °C.

Parameter	Units	Enriched PAO culture Smolders *et al.* (1994)	Enriched GAO cultures				
			Anaerobic model Zeng *et al.* (2003)	Liu *et al.* (1994)	Filipe *et al.* (2001a)	Zeng *et al.* (2003)	This study
SRT	d	8.0	7.0	7.5 - 8.0	7.0	6.6	10.0
pH		7.0	7.0	7.0 - 8.0	7.0	7.0	7.0
Acetate consumed	C-mol	-1.00	-1.00	-1.00	-1.00	-1.00	-1.00
Gly/HAc	C-mol/C-mol	-0.50	-1.12	-1.11 - -1.25	-0.83	-1.20	-1.20 ± 0.19
PHA/HAc	C-mol/C-mol	1.33	1.85	1.76 - 1.91	1.65	1.91	1.97 ± 0.13
PHB/HAc	C-mol/C-mol	1.21	1.36	1.17	1.26	1.39	1.28 ± 0.06
PHV/HAc	C-mol/C-mol	0.12	0.46	0.68	0.38	0.52	0.69 ± 0.07
PHV:PHB ratio	C-mol/C-mol	0.10	0.34	0.58	0.31	0.38	0.54 ± 0.04

Figure 2.3 presents the stoichiometric ratios measured during the short-term batch tests performed from 10 to 40 °C. Between 15 and 35 °C, stoichiometric ratios were essentially constant. In addition, average stoichiometric values (1.36 ± 0.16 Gly/HAc, 1.32 ± 0.07 PHB/HAc; 0.60 ± 0.06 PHV/HAc; and, 0.45 ± 0.05 PHV/PHB, in C-mol/C-mol units) are within the ranges observed in other reports (see Table 2.1). However, at 10 °C, the observed ratios were significantly lower (in some cases even 70 %) than the stoichiometric values measured in the rest of the temperature interval. 40 °C seemed to cause a complete collapse of activity since no acetate uptake was observed at this temperature.

Based on results presented on Figure 2.3, it can be concluded that GAO anaerobic stoichiometry is insensitive to temperature changes from 15 to 35 °C. However, it is significantly affected at low temperature (10 °C) meanwhile 40 °C was above the maximal temperature where GAO were able to survive.

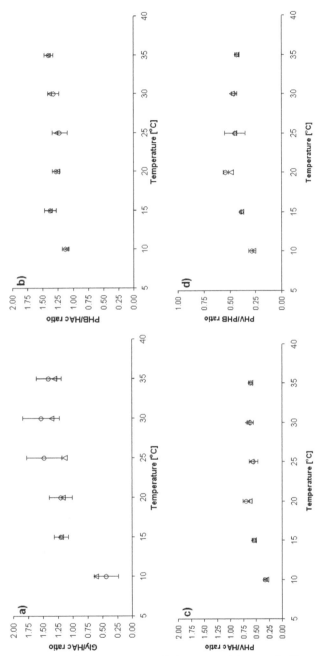

Figure 2.3. Stoichiometric parameters of the anaerobic metabolism of Glycogen Accumulating Organisms as function of temperature: **a)** Glycogen/Acetate ratio; **b)** PHB/Acetate ratio; **c)** PHV/Acetate ratio; and, **d)** PHV/PHB ratio. Error bars indicate standard deviations of measurements.

2.3.3. Maximum anaerobic acetate uptake of GAO

The maximum anaerobic acetate uptake rate ($q_{SA,GAO}^{MAX}$) of GAO at the different studied temperatures is presented in Figure 2.4. $q_{SA,GAO}^{MAX}$ at 20 °C (0.20 C-mol HAc/C-mol biomass/h) is in the range of previous observed rates reported by Zeng *et al.* (2003) and Filipe *et al.* (2001a) (0.16 - 0.18 and 0.24 C-mol/C-mol biomass/h, respectively). However, these values differ from the acetate uptake rates measured by Sudiana *et al.* (1999), Liu *et al.* (1997) and Schuler and Jenkins (2003) who reported rates of about 0.04 - 0.08 C-mol/C-mol biomass/h. A direct explanation to this difference cannot be found on the basis of the reported data but a possible reason on the different pH applied. According to Filipe *et al.* (2001a) and Schuler and Jenkins (2002), $q_{SA,GAO}^{MAX}$ decreases as pH increases. Therefore, it seems that a lower pH value applied in this study (pH 7.0 ± 0.05) compared to the higher pH used by Schuler and Jenkins (2003) (from pH 7.15 to 7.25) and by Sudiana *et al.* (1999) and Liu *et al.* (1997) (pH 7.0 - 8.0) helps to partially explain the higher $q_{SA,GAO}^{MAX}$ observed in this research. Filipe *et al.* (2001a) and Zeng *et al.* (2003) used similar pH levels (pH 6.8 - 7.1 and 6.85 – 7.05, respectively) and also observed higher rates. Nevertheless, due to the considerable difference in acetate uptake rates for a small pH variation it can be concluded that pH cannot be the sole explanation to this result.

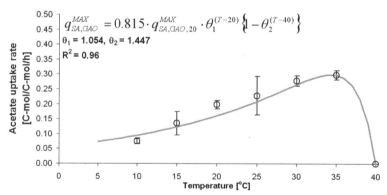

Figure 2.4. Effect of temperature on the maximum acetate uptake rate of Glycogen Accumulating Organisms. Error bars indicate standard deviations of measurements.

$q_{SA,GAO}^{MAX}$ increased as temperature rose from 10 to 35 °C and, at 35 °C, the optimal maximum acetate uptake was observed (0.30 C-mol HAc/C-mol biomass/h). GAO activity experienced a sudden declination above 35 °C and at 40 °C acetate uptake ceased. $q_{SA,GAO}^{MAX}$ was satisfactorily described (R^2 = 0.96) using a double Arrhenius expression (θ_1 = 1.054 and θ_2 = 1.447).

2.3.4. Anaerobic ATP maintenance coefficient

During anaerobic batch tests to determine $m_{ATP,GAO}^{an}$, the glycogen depletion profile did not level off and, moreover, it was not depleted in any experiment (data not shown). Figure 2.5 shows the effect of temperature on the specific anaerobic maintenance coefficient of GAO, $m_{ATP,GAO}^{an}$. At 20 °C, a $m_{ATP,GAO}^{an}$ of 3.3 ± 0.20 x 10^{-3} mol ATP/C-mol Biomass/h was observed which was slightly higher than the values reported for other GAO cultures (2.4 to 2.7 x 10^{-3}) (Filipe *et al.*, 2001a; Zeng *et al.*, 2003).

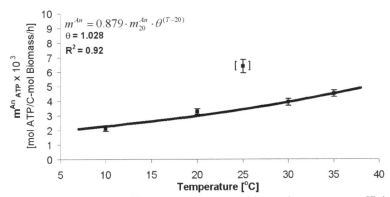

Figure 2.5. Temperature effects on the anaerobic maintenance coefficient of Glycogen Accumulating Organisms. Error bars indicate standard deviations of measurements.

Due to unknown reasons, $m_{ATP,GAO}^{an}$ at 25 °C clearly differs from the trendline of the rest of maintenance coefficients observed at 10, 20, 30 and 35 °C. Therefore it was not considered to calculate the simplified Arrhenius expression (Figure 2.5). The 40 °C maintenance value was also discarded because of, as previously mentioned, the complete collapse of activity. The temperature effects on maintenance coefficients (from 10 to 35 °C) are well predicted with a simplified Arrhenius equation (factor = 0.879, θ = 1.028, R^2 = 0.92).

A comparison between the temperature effects on anaerobic maintenance requirements of PAO and GAO as well as other coefficients for PAO and GAO cultures reported by other authors are shown in Figure 2.6.

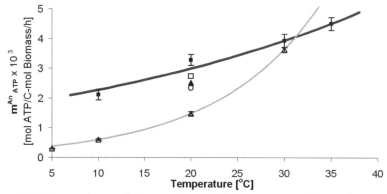

Figure 2.6. Comparison of temperature effects on anaerobic maintenance of Polyphosphate Accumulating Organisms (gray line: trendline, $\triangle$: Brdjanovic *et al.*, 1997; $\blacktriangle$: Smolders *et al.*, 1994) and Glycogen Accumulating Organisms (black line: trendline; $\blacksquare$: this study; $\square$: Filipe *et al.*, 2001a; $\circ$: Zeng *et al.*, 2003).

Despite the slightly higher $m^{an}_{ATP,GAO}$ measured in the present study when the temperature dependency of $m^{an}_{ATP,PAO}$ (Brdjanovic *et al.*, 1997) and $m^{an}_{ATP,GAO}$ (this study) are compared (Figure 2.6) two different trends are observed. It seems that especially at low temperature (below 30 $^{\circ}$C) the anaerobic maintenance requirements of PAO are lower than those of GAO whereas, based on extrapolation of the temperature dependency of PAO, the opposite occurs as temperature increases above 30 $^{\circ}$C (Figure 2.6).

2.3.5. Performance of PAO SBR

The PAO SBR was continuously operated for more than 90 days. It reached steady-state conditions in the first 50 days of operation. Cycle measurements were carried out, after the SBR reached steady-state conditions, to determine the biomass activity. A typical cycle profile of a SBR enriched with PAO is presented in Figure 2.7. It exhibits anaerobic HAc consumption, significant anaerobic P-release (PO_4^{3-}-P), PHA storage (as PHB and PHV) and glycogen consumption. Under aerobic conditions, full-aerobic P-uptake, PHA oxidation and glycogen production were observed. Average MLSS and MLVSS concentrations at the end of aerobic phase were 2980 and 2105 mg/L, respectively, resulting in an average MLVSS/MLSS ratio of 0.71. The low observed MLVSS/MLSS ratio indicated high storage of inorganic matter, presumably poly-P. Anaerobic P-released/HAc uptake (P/HAc) ratio was 0.39 PO_4^{3-}-P mol/HAc C-mol. Observed biomass activity showed the typical pattern of lab-scale systems enriched with PAO (Smolders *et al.*,

1994a; Brdjanovic *et al.*, 1997, 1998a; Filipe *et al.*, 2001c). Furthermore, Figure 2.8 presents a FISH analysis of the population distribution in PAO SBR showing that PAO were the dominant organisms. The dominance of PAO was confirmed by a quantification of the population distribution shown in Figure 2.8, which showed that PAO comprised about 85 % of the total population, GAO 12 % and other bacteria 3 %.

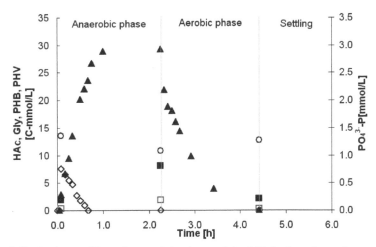

Figure 2.7. Cycle profiles observed in the enriched Polyphosphate Accumulating Organisms culture: acetate (◊), glycogen (○), PHB (■), PHV (□) and phosphate (▲).

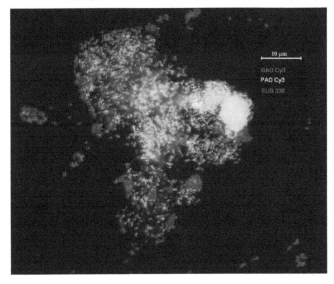

Figure 2.8. Bacterial population distribution in the enriched culture of Polyphosphate Accumulating Organisms by applying Fluorescence in situ Hybridization techniques (bar indicates 10 µm; blue: Eubacteria, red: Glycogen accumulating organisms and green: Polyphosphate-accumulating organisms).

2.3.6. Maximum anaerobic acetate uptake of PAO and GAO cultures

Brdjanovic *et al.* (1997, 1998a) studied in detail the effects of temperature on PAO from 5 to 30 °C. In the present study, the short-term temperature effects on the maximum acetate uptake rate of PAO ($q_{SA,PAO}^{MAX}$) were studied from 20 to 40 °C in order to compare the temperature effects on PAO and GAO. Figure 2.9 displays, simultaneously, the acetate uptake rates of PAO observed by Brdjanovic *et al.*, (1997) and the rates measured in the present study for the PAO culture. Additionally, the acetate uptake rates of GAO are included to compare the temperature effects on the rates of both types of microorganisms.

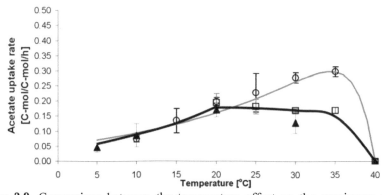

Figure 2.9. Comparison between the temperature effect on the maximum acetate uptake rates of Polyphosphate Accumulating Organisms (black line: trendline; ■: Brdjanovic *et al.*, 1997; □: this study) and Glycogen Accumulating Organisms (gray line: trendline; ○: this study).

For the PAO culture, the $q_{SA,PAO}^{MAX}$ observed in this study at 20 °C (0.17 C-mol HAc/C-mol biomass/h) was similar to the rates observed by previous authors (0.17-0.19 C-mmol/C-mmol Biomass/h) (Brdjanovic *et al.*, 1997; 1998a; Filipe *et al.*, 2001c). However, it was significantly lower than the value reported by Smolders *et al.* (1994a) (0.43 C-mol HAc/C-mol biomass/h). From 20 to 35 °C, $q_{SA,PAO}^{MAX}$ remained constant (0.17 C-mol HAc/C-mol biomass/h) which agrees with previous research (Brdjanovic *et al.*, 1997). As observed previously with GAO, PAO did not show activity at 40 °C. $q_{SA,PAO}^{MAX}$ profile could not be fitted to a double Arrhenius expression (Equation 2.3) as GAO acetate uptake kinetics. PAO did not show a sudden activity declination immediately after reaching their optimal temperature level. On the opposite, above 20 °C and up to 35 °C, the kinetic rate remained constant and declined between 35 and 40 °C. The double Arrhenius expression could not be applied to $q_{SA,PAO}^{MAX}$ profile because of the constant range observed from 20 to 35 °C.

2.4. Discussion

2.4.1. Enrichment of the GAO culture

Although phosphorus concentration in the influent was reduced to a minimum level to suppress PAO growth (Liu *et al.*, 1997), according to a quantification of the population distribution (Figure 2.2) GAO and PAO comprised around 75 % and 20 % of total microbial population, respectively. Based on the experimental observations, PAO population seemed larger than expected. This means that about 20 % of the total C-source (as acetate) fed to the GAO SBR might have been consumed by PAO and, according to Smolders *et al.* (1994a), should have led to an approximate net P-release of 18 mg/L and an observed P/HAc ratio of 0.10 P-mol/HAc C-mol. However, after GAO SBR reached steady-state conditions, the observed P-release remained always below 3 mg/L. Therefore, the contribution of PAO to the observed biomass activity was negligible based on the low P/HAc ratio measured under anaerobic conditions (0.01) during the whole experimental phase. The latter indicates that PAO, due to the low influent P-concentration, did not store poly-P (as the high MLVSS/MLSS ratio indicates) and, thus, their activity was effectively suppressed. Furthermore, it has been observed that PAO are not able to take up acetate when their poly-P content is depleted (Brdjanovic *et al.*, 1998b). Nevertheless, due to unclear reasons the bacterial population quantification indicated that PAO were not totally washed out but, according to SBR performance, remained inactive in the SBR. These observations indicate that there was a discrepancy between the FISH technique and the experimental data. Therefore, since the experimental results of this study are within the ranges of previous reported values (Table 2.1) it may be concluded that the results of the bacterial population quantification of the GAO SBR, via the FISH image (Figure 2.2), do not reflect the typical microbial population during the experimental period. Deviations resulting from FISH analysis might have led to an overestimation of the PAO population. It could not be discarded that FISH probes were not specific enough to target the actual bacterial populations.

2.4.2. Temperature effects on anaerobic conversions

In the present work, the observed anaerobic stoichiometric ratios of GAO were in the range of previous reported values (Table 2.1) and remained insensitive to temperature changes from 15 to 35 °C (Figure 2.3). However, temperature had an important effect on GAO stochiometry at 10 and 40 °C.

Stoichiometric ratios at 10 °C were lower than ratios measured in the rest of the studied temperature interval (Gly/HAc ratio was even 70 % lower). To some extent it seemed that, at 10 °C, GAO suffered a partial inactivation, inhibition or modification of their metabolic pathways that led to lower glycogen consumption (0.43 C-mo/C-mol), and PHB and PHV production per acetate consumed (1.13 and 0.32 C-mol/C-mol, respectively). Zeng *et al.* (2002) proposed that, for PAO, the required ATP (0.50 + α) for acetate transport is produced by glycogen hydrolysis ($1+2\alpha$ C-mol Gly/C-mol HAc). Where α is the energy necessary for HAc transport through the cell membrane. At pH 7.0 and 20 °C, α is approximately 0.06 mol ATP/C-mol HAc (Filipe *et al.*, 2001a) which leads to 0.56 mol ATP/C-mol HAc and 1.12 C-mol Gly/C-mol HAc. In consequence, it is not clear how GAO were able to produce the required ATP to cover their energy requirements at 10 °C. Even if it is assumed that no energy is needed for HAc transport through cell membrane ($\alpha = 0$) 1 C-mol Gly/C-mol HAc would still be needed. Similar results are obtained when another GAO model is used (Filipe *et al.*, 2001a). Therefore, ATP might have been produced through another metabolic pathway where glycogen and poly-P are not involved. In consequence, no firm conclusions can be drawn regarding the rest of the ratios measured at 10 °C. Further research is needed to clarify the temperature effect on GAO at 10 °C. But, it should be noticed that PAO's anaerobic stoichiometry remained stable at 10 and even 5 °C (Brdjanovic *et al.*, 1997).

No acetate uptake was observed at 40 °C. Based on the $q_{SA,GAO}^{MAX}$ profile (Figure 2.4), it can be stated that an irreversible enzymatic inactivation caused by exposition to high temperature may have been the main cause of activity collapse (Bazin and Prosser, 2000).

It can be concluded that GAO anaerobic stoichiometry, as reported before for PAO (Brdjanovic *et al.*, 1997), is insensitive to short-term (hours) temperature changes in a broad temperature range (from 15 to 35 °C). At 10 °C, GAO cultures have a clear shift in their metabolism whereas PAO still have their normal conversions. At 40 °C, both microorganisms suffered a collapse of activity.

Regarding the anaerobic kinetics, GAO anaerobic acetate uptake rates (from 10 to 40 °C) showed a moderate temperature dependency. The first Arrhenius coefficient θ_1 (1.054) is lower than the coefficient found by Brdjanovic *et al.* (1997) for PAO cultures ($\theta = 1.078$) and coefficients suggested for nitrification and fermentation processes (1.120 and 1.070, respectively) (Henze *et al.*, 2000).

It is suggested to take into account the found temperature dependencies of PAO and GAO, through their corresponding Arrhenius coefficients, to extend the applicability range of current activated sludge models (Henze *et al.*, 2000) to a wider temperature range.

2.4.3. Anaerobic maintenance coefficient

A slightly higher $m^{an}_{ATP,GAO}$ was found in this study at 20 °C than in other reports (Filipe *et al.*, 2001a; Zeng *et al.*, 2003). Nevertheless, Figure 2.6 shows that, at 20 °C, the predicted value of the trendline that describes the temperature dependency of the anaerobic maintenance coefficient of GAO is in the range of previous reported coefficients.

$m^{an}_{ATP,GAO}$ at 20 °C was 36% higher than most of the values reported in the literature for PAO cultures (1.47 to 2.5 x 10^{-3}) (Smolders *et al.*, 1995; Brdjanovic *et al.*, 1997; Filipe *et al.*, 2001c). This higher $m^{an}_{ATP,GAO}$ partially explains the results of Matsuo (1994) who observed better P-removal performance when the anaerobic SRT was increased. It is possible that a longer anaerobic SRT created favorable conditions for PAO because of the higher anaerobic maintenance requirements of GAO.

2.4.4. Competition between PAO and GAO cultures

There was a clear difference between the effects of temperature on the anaerobic metabolism of PAO and GAO. Figure 2.9 shows that both microorganisms present similar maximum anaerobic acetate uptake rates below 20 °C. At higher temperature (> 20 °C), $q^{MAX}_{SA,PAO}$ is insensitive to temperature changes (0.17 C-mol HAc/C-mol Biomass/h) and remains constant up to 35 °C. Meanwhile $q^{MAX}_{SA,GAO}$ increases from 20 °C and reaches an optimal value around 35 °C (0.30 C-mol HAc/C-mol Biomass/h) which is almost two times the $q^{MAX}_{SA,PAO}$ value.

The energy source for acetate uptake seems to be the main metabolic difference between PAO and GAO. For PAO, poly-P hydrolysis produces the required energy for acetate uptake while GAO obtain the energy from glycogen hydrolysis. Thereupon, poly-P hydrolysis could be the limiting pathway of PAO's metabolism above 20 °C.

Temperature effects on anaerobic maintenance requirements of PAO and GAO (Figure 2.6) might be also a decisive factor in the PAO-GAO competition. In the present study, similar acetate uptake rates for PAO and GAO were found below 20 °C. Additionally, comparable aerobic yields and aerobic maintenance requirements have been reported for both microorganisms (Filipe *et al.*, 2001b; Zeng *et al.*, 2003). These findings indicate that PAO and GAO might coexist at medium and lower temperature ranges ($\leq$ 20 $^{\circ}$C) and this could be detrimental for the EBPR process. However, successful EBPR performance has been reported at temperatures below 20 $^{\circ}$C in full-scale wwtp (van Veldhuizen *et al.*, 1999; Ybstebɸ *et al.*, 2000; Meijer *et al.*, 2002b; Tykesson *et al.*, 2005). The latter could be partially explained based on the higher maintenance requirements of GAO observed in the present study at lower temperature. Higher maintenance requirements lead to higher energy consumption and lower net yield coefficients, therefore since GAO spent more energy on maintenance lower net yield coefficients of GAO than PAO are expected at temperature below 20 $^{\circ}$C.

Higher acetate uptake rates observed at higher temperature ($>$ 20 $^{\circ}$C) and a possibly inhibition at lower temperature (around and below 10 $^{\circ}$C) indicate that GAO appear to be mesophilic microorganisms. On the other hand, acetate uptake rate limitation at higher temperature ($>$20 $^{\circ}$C) might suggest that PAO could be considered psychrophilic.

In general, the effects of temperature regarding the maximum acetate uptake rates and anaerobic maintenance coefficients found in the present study help to explain the proliferation of GAO at higher temperature and consequently the EBPR instability reported in wwtp when treating warm effluents. However, there might be more factors than competition for substrate and energy maintenance requirements that limit the proliferation of GAO at full-scale plants. How GAO metabolism is affected by environmental and operating conditions may play a major role on their appearance, coexistence or absence at full-scale treatment plants at low and moderate temperatures ($\leq$ 20 $^{\circ}$C). Besides, at 20 $^{\circ}$C, the usual applied temperature for lab-scale plants, the PAO-GAO competition takes place in a narrow range due to their low metabolic differences. Therefore, the possible impact of the inoculum and wastewater origin, that determines the initial PAO-GAO populations, might have a significant influence on their competition as observed by Matsuo *et al.* (1982) (cited by Satoh *et al.*, 1994).

The findings of the present study are important to get a better understanding about the interaction between PAO and GAO. However, since short-term temperature tests are performed for a few hours neither changes nor adaptations of bacterial populations can be expected. Long-term temperature effect tests should be carried out to account for potential population changes and adaptations. Brdjanovic *et al.* (1997, 1998) found similar temperature dependencies for PAO from short- and long-term temperature tests. Thus, likely, similar results from short-term experiments would be obtained at long-term for GAO. In addition, long-term experiments could help to elucidate unexplained observations that remained in this research.

2.5. Conclusions

The effects of temperature on Polyphosphate Accumulating Organisms (PAO) and Glycogen Accumulating Organisms (GAO) cultures indicate that GAO have clear advantages over PAO for substrate uptake at temperature higher than 20 °C. Below 20 °C, maximum acetate uptake rates of both microorganisms were similar. However, lower maintenance requirements at temperature lower than 30 °C give PAO metabolic advantages in the PAO-GAO competition. In consequence, Polyphosphate Accumulating Organisms could be considered to be psychrophilic microorganisms while Glycogen Accumulating Organisms appear to be mesophilic. These findings contribute to understand and support the practical observations related to the stability of the Enhanced Biological Phosphorus Removal process at wastewater treatment plants operated under cold weather conditions. They also explain the proliferation of Glycogen Accumulating Organisms, which leads to instability of the biological phosphorus removal process, in treatment plants from hot regions or treating high-temperature industrial effluents. It is suggested to consider the observed temperature dependencies of PAO and GAO in order to extend the applicability of current activated sludge models to a wider temperature range.

Acknowledgements

The authors acknowledge the National Council for Science and Technology (CONACYT, Mexico) and the Autonomous University of the State of Mexico for the scholarship awarded to Carlos Manuel Lopez Vazquez. The useful help provided by Joao Xavier for the population quantifications by the FISH images is greatly acknowledged. Special thanks to the lab-staff from UNESCO-IHE Institute for Water Education.

References

Amann R. I. (1995). *In situ* identication of microorganisms by whole cell hybridization with rRNA-targeted nucleic acid probes. In: Akkermans ADL, van Elsas JD, de Bruijn FJ, editors. Molecular Microbial Ecology Manual. London: Kluwer Academic Publisher. p 1-15.

APHA (1998) Standard methods for the examination of water and wastewater. Washington, DC: 20th ed. American Public Health Association/American Water Works Association/Water Environment Federation.

Barnard JL, Steichen MT (2006) Where is nutrient removal going on? Water Sci Technol 53(3):155-164.

Bazin MJ, Prosser JI (2000) Physiological models in microbiology. CRC Series in mathematical models in microbiology. Florida: CRC Press.

Brdjanovic D, van Loosdrecht MCM, Hooijmans CM, Alaerts GJ, Heijnen JJ (1997) Temperature effects on physiology of biological phosphorus removal. ASCE J Environ Eng 123 (2):144-154.

Brdjanovic D, Logemann S, van Loosdrecht MCM, Hooijmans CM, Alaerts GJ, Heijnen JJ (1998a) Influence of temperature on biological phosphorus removal: process and molecular ecological studies. Water Res 32 (4):1035-1048.

Brdjanovic D, van Loosdrecht MCM, Hooijmans CM, Mino T, Alaerts G, Heijnen JJ (1998b) Effect of polyphosphate limitation on phosphorus-accumulating micro-organisms. J Appl Microbiol Biotechnol 50:273-276.

Cech JS, Hartman P (1993) Competition between phosphate and polysaccharide accumulating bacteria in enhanced biological phosphorus removal systems. Water Res. 27:1219-1225.

Crocetti GR, Hugenholtz P, Bond PL, Schuler A, Keller J, Jenkins D, Blackall LL. (2000) Identification of polyphosphate-accumulating organisms and design of 16S rRNA-directed probes for their detection and quantitation. Appl Environ Microbiol 66(3):1175–1182.

Crocetti GR, Banfield JF, Keller J, Bond PL, Blackall LL (2002) Glycogen accumulating organisms in laboratory-scale and full-scale wastewater treatment processes. Microbiol 148:3353-3364.

Erdal UG, Erdal ZK, Randall CW (2003) The competition between PAO (phosphorus accumulating organisms) and GAO (glycogen accumulating organisms) in EBPR (enhanced biological phosphorus removal) systems at different temperatures and the effects on system performance. Water Sci Technol 47(11):1-8.

Filipe CDM, Daigger GT, Grady Jr CPL (2001a) A metabolic model for acetate uptake under anaerobic conditions by glycogen-accumulating organisms: stoichiometry, kinetics and effect of pH. Biotechnol Bioeng 76(1):17-31.

Filipe CDM, Daigger GT, Grady Jr CPL (2001b) Effects of pH on the aerobic metabolism of phosphate-accumulating organisms and glycogen-accumulating organisms. Water Environ Res 73(2):213-222.

Filipe CDM, Daigger GT, Grady Jr CPL (2001c) Stoichiometry and kinetics of acetate uptake under anaerobic conditions by an enriched culture of phosphorus-accumulating organisms at different pHs. Biotechnol Bioeng 76(1):32-43.

Grady Jr. CPL, Filipe CDM (2000) Ecological engineering from biorreactors for wastewater treatment. Water Air Soil Pollut 123:117-132.

Henze M, Gujer W, Mino T, van Loosdrecht MCM (2000) Activated sludge models ASM1, ASM2, ASM2d and ASM3. IWA Scientific and Technical Report No. 9. IWA Task Group on Mathematical Modelling for Design and Operation of Biological Wastewater Treatment. London: IWA Publishing.

Jobbàgy A, Litherathy B, Tardy G (2002) Implementation of glycogen accumulating bacteria in treating nutrient-deficient wastewater. Water Sci Technol 46(1-2):185-190.

Liu WT, Mino T, Nakamura K, Matsuo T (1994) Role of glycogen in acetate uptake and polyhydroxyalkanoate synthesis in anaerobic-aerobic activated sludge with a minimized polyphosphate content. J Ferment Bioeng 77(5):535-540.

Liu WT, Mino T, Nakamura K, Matsuo T (1996) Glycogen accumulating population and its anaerobic substrate uptake in anaerobic-aerobic activated sludge without biological phosphorus removal. Water Res 30(1):75-82.

Liu WT, Nakamura K, Matsuo T, Mino T (1997) Internal energy-based competition between polyphosphate- and glycogen-accumulating bacteria in biological phosphorus removal reactors-effect of P/C feeding ratio. Water Res 31(6):1430-1438.

Lopez C, Pons MN, Morgenroth E (2006) Endogenous process during long-term starvation in activated sludge performing enhanced biological phosphorus removal. Water Res 40:1519-1530.

Matsuo Y, Kitagawa M, Tanaka T, Miya A (1982) Sewage and night-soil treatment by anaerobic-aerobic activated sludge processes. Proceedings Environ Sanitary Eng Res 19:82-87 (in Japanese).

Matsuo Y (1994) Effect of the anaerobic solids retention time on enhanced biological phosphorus removal. Water Sci Technol 30(6):193-202.

Meijer SCF, van Loosdrecht MCM, Heijnen JJ (2002b) Modelling the start-up of a full-scale biological nitrogen and phosphorus removing WWTP. Water Res 36(19):4667-4682.

Mino T, van Loosdrecht MCM, Heijnen JJ (1998) Microbiology and biochemistry of the enhanced biological phosphorus removal process. Water Res 32(11):3193-3207.

Oehmen A, Yuan Z, Blackall LL, Keller J (2005) Comparison of acetate and propionate uptake by polyphosphate accumulating organisms and glycogen accumulating organisms. Biotechnol Bioeng 91(2): 162-168.

Otsu N (1979) A threshold selection method from gray-level histograms. IEEE Trans Sys Man Cyber 9 (1): 62-66.

Panswad T, Doungchai A, Anotai J (2003) Temperature effect on microbial community of enhanced biological phosphorus removal system. Water Res 37:409-415.

Park JK, Ahn CH, Whang LM, Lee EK, Lee YO, Probst T, Reichardt RN, Brauer J W (2001) Problems in biological phosphorus removal with dairy wastewater. Proceedings of the 7th Annual Industrial Wastes Technical and Regulatory Conference. Water Environment Federation. August 12–15, 2001, Charleston.

Satoh H, Mino T, Matsuo T (1994) Deterioration of enhanced biological phosphorus removal by the domination of microorganisms without polyphosphate accumulation. Water Science and Technology, 30(6):203-211.

Saunders AM, Oehmen A, Blackall LL, Yuan Z, Keller J (2003) The effect of GAO (glycogen accumulating organisms) on anaerobic carbon requirements in full-scale Australian EBPR (enhanced biological phosphorus removal) plants. Water Sci Technol 47(11):37-43.

Schuler AJ, Jenkins D (2002) Effects of pH on enhanced biological phosphorus removal metabolisms. Wat Sci Technol 46(4-5):171-178.

Schuler AJ, Jenkins D (2003) Enhanced biological phosphorus removal from wastewater by biomass with different phosphorus contents, Part 1: Experimental results and comparison with metabolic models. Water Environ Res 75(6):485-498.

Smolders GJF, Klop JM, van Loosdrecht MCM, Heijnen JJ (1995) A metabolic model of the biological phosphorus removal process: I. Effect of the sludge retention time. Biotechnol Bioeng 48(3):222-233.

Smolders GJF, van Der Meij J, van Loosdrecht MCM, Heijnen JJ (1994a) Model of the anaerobic metabolism of the biological phosphorus removal processes: stoichiometry and pH influence. Biotechnol Bioeng 43:461-470.

Smolders GJF, van Der Meij J, van Loosdrecht MCM, Heijnen JJ (1994b) Stoichiometric model of the aerobic metabolism of the biological phosphorus removal process. Biotechnol Bioeng 44:837-848.

Sudiana IM, Mino T, Satoh H, Nakamura K, Matsuo T (1999) Metabolism of enhanced biological phosphorus removal and non-enhanced biological phosphorus removal sludge with acetate and glucose as carbon source. Wat Sci Tech 39(6):29-35.

Thomas M, Wright P, Blackall LL, Urbain V, Keller J (2003) Optimization of Noosa BNR plant to improve performance and reduce operating costs. Water Sci Technol 47(12):141-148.

Tykesson E, Jönsson L-E, la Cour Jansen J (2005) Experience from 10 years of full-scale operation with enhanced biological phosphorus removal. Water Sci Technol 52(12):151-159.

van Loosdrecht MCM, Henze M (1999) Maintenance, endogenous respiration, lysis, decay and predation. Water Sci Technol 39(1):107-117.

van Veldhuizen HM, van Loosdrecht MCM, Heijnen JJ (1999) Modelling biological phosphorus and nitrogen removal in a full scale activated sludge process. Water Res 33(16):3459-3468.

Whang LM, Park JK (2002) Competition between polyphosphate- and glycogen-accumulating organisms in biological phosphorus removal systems- effect of temperature. Water Sci Technol 46(1-2):191-194.

Whang LM, Park JK (2006) Competition between polyphosphate- and glycogen-accumulating organisms in enhanced biological phosphorus removal systems: effect of temperature and sludge age. Water Environ Res 78(1):4-11.

Wong MT, Mino T, Seviour RJ, Onuki M, Liu WT (2005) *In situ* identification and characterization of the microbial community structure of full-scale enhanced biological phosphorous removal plants in Japan.Water Res 39(13):2901-2914.

Ydstebф L, Bilstad T, Barnard J (2000) Experience with biological nutrient removal at low temperatures. Water Environ Res 72(4):444-454.

Zeng RJ, van Loosdrecht MCM, Yuan Z, Keller J (2002) Proposed modifications to metabolic model for glycogen accumulating organisms under anaerobic conditions. Biotechnol Bioeng 80(3):277-279.

Zeng RJ, van Loosdrecht MCM, Yuan Z, Keller J (2003) Metabolic model for glycogen-accumulating organisms in anaerobic/aerobic activated sludge systems. Biotechnol Bioeng 81(1):92-105.

3
Chapter

Temperature effects on the aerobic metabolism of GAO

Content

This chapter has been published as:

Lopez-Vazquez CM, Song YI, Hooijmans CM, Brdjanovic D, Moussa MS, Gijzen HJ, van Loosdrecht MCM (2008) Temperature effects on the aerobic metabolism of glycogen accumulating organisms. *Biotech Bioeng* 101(2):295-306.

Abstract

Short-term temperature effects on the aerobic metabolism of GAO were investigated within a temperature range from 10 to 40 °C. *Candidatus Competibacter Phosphatis*, known GAO, were the dominant microorganisms in the enriched culture comprising 93 ± 1% of total bacterial population as indicated by Fluorescence *in situ* Hybridization (FISH) analysis. Between 10 and 30 °C, the aerobic stoichiometry of GAO was insensitive to temperature changes. Around 30 °C, the optimal temperature for most of the aerobic kinetic rates was found. At temperatures higher than 30 °C, a decrease on the aerobic stoichiometric yields combined with an increase on the aerobic maintenance requirements were observed. An optimal overall temperature for both anaerobic and aerobic metabolisms of GAO appears to be found around 30 °C. Furthermore, within a temperature range (10 to 30 °C) that covers the operating temperature range of most of domestic wastewater treatment systems, GAO's aerobic kinetic rates exhibited a medium degree of dependency on temperature (θ = 1.046 to 1.082) comparable to that of PAO. We conclude that GAO do not have metabolic advantages over PAO concerning the effects of temperature on their aerobic metabolism, and competitive advantages are due to anaerobic processes.

3.1. Introduction

Among several studies regarding the influence of environmental and operating conditions on the EBPR process, different authors have underlined temperature as one of the determinant factors to understand the competition between PAO and GAO (Panswad *et al.*, 2003; Erdal *et al.*, 2003; Whang and Park, 2006; Lopez-Vazquez *et al.*, 2007). These researchers agreed that temperature has a major influence on the maximum anaerobic substrate uptake rates of these microorganisms providing important insights to improve the understanding of the PAO-GAO interaction and competition.

Despite the fact that biomass production and glycogen storage take place under aerobic conditions, limited attention has been paid to the effects of temperature on the aerobic metabolism of GAO as opposite to PAO where Brdjanovic *et al.* (1997, 1998) performed a systematic study concerning the temperature effects on the biological phosphorus removal. In that study, they were able to describe the temperature dependencies of the different processes involved in PAO's aerobic metabolism from 5 to 30 °C. A similar systematic study regarding the temperature effects on the different processes involved in the aerobic metabolism of a GAO culture has not been reported yet. Such a study could provide important information to understand the occurrence of these microorganisms at full-scale wastewater treatment plants (WWTP). Furthermore, the aerobic temperature dependencies of GAO could be combined with their anaerobic temperature dependencies (Lopez-Vazquez *et al.*, 2007) to model the interaction between PAO and GAO at different temperatures. This may furthermore lead to better understanding of the PAO-GAO competition helping to comprehend the stability of the EBPR process at different temperature. Therefore, there is a clear need for studying and determining the temperature dependencies of the aerobic metabolism of GAO.

The temperature effects on (1) the aerobic stoichiometry and (2) the aerobic kinetics of GAO were evaluated using a lab-enriched culture. Since the main objective was to investigate the effects on the metabolism of these microorganisms, short-term temperature changes on sludge cultivated at 20 °C were studied. The research was carried out by executing aerobic batch tests within a temperature range from 10 to 40 °C that covers the operating temperature interval of most of domestic and industrial activated sludge wastewater treatment plants.

3.2. Materials and methods

3.2.1. Continuous operation of the sequencing batch reactor

A GAO culture was enriched in a double-jacketed lab-scale sequencing batch reactor (SBR) as described in chapter 2 (section 2.2.1). The SBR had a working volume of 2.5 L. Activated sludge from a domestic WWTP (Nieuwe Waterweg, Hoek van Holland, The Netherlands) was used as inoculum. The SBR was operated at 20 ± 0.5 °C in cycles of 6 hours (2.25 h anaerobic, 2.25 h aerobic and 1.5 h settling phase). pH was maintained at 7.0 ± 0.1 by dosing 0.3 M HCl and 0.2 M NaOH. At the beginning of the cycle, 1.25 L of synthetic medium was fed to the SBRs over a period of 5 minutes. The SBR was operated with a solids retention time (SRT) of 10 days and a hydraulic retention time (HRT) of 12 h. Further details regarding the operating conditions of the SBR can be found elsewhere (section 2.2.1. of this thesis).

The aerobic batch experiments at different temperatures were performed in a separate reactor after the biomass activity reached steady-state conditions. Thus, the SBR was only used as a source of enriched GAO biomass.

3.2.2. Operation of the batch reactor

A double-jacketed laboratory fermenter with a maximal volume of 0.5 L was used for the execution of the aerobic batch experiments. The experiments were performed at controlled temperatures and pH (7.00 ± 0.05). pH was maintained by dosing 0.2 M HCl and 0.2 M NaOH. At the beginning of each experiment, enriched GAO sludge was manually transferred from the parent SBR to the batch reactor. The amount of transferred sludge was 450 mL, which was higher than the amount of sludge wasted daily from the SBR. In order to avoid major disturbances on the continuous operation and steady-state conditions of the parent SBR, the waste of sludge was stopped for almost 2 days after the execution of any of the aerobic batch tests to compensate for the sludge transfer. The sludge used in the batch tests was not returned to the SBR. A double-jacketed respirometer with a working volume of 10 mL was connected to the batch reactor for the online measurement of the oxygen uptake rate (OUR). Due to the fast respirometric activity observed within the first minutes of the aerobic batch test, sludge was pumped (circulated) from the batch reactor to the respirometer in cycles of 3 minutes in the first 30 - 40 minutes (1 min pumping then the recirculation was stopped and the oxygen depletion was measured for 2 min) and in cycles of 6 minutes during the rest of the assay (1 min pumping and the oxygen depletion was measured for 5 min) as described by Brdjanovic *et al.* (1997). For the determination of the oxygen requirements for aerobic maintenance processes, the length of the cycle was 25 min (5 min pumping per cycle and 20 min recording the oxygen depletion). Measured data were continuously stored in a computer. Minimum oxygen concentration in the double-jacketed respirometer was always kept above 2 mg/L.

3.2.3. Synthetic media

Synthetic media supplied to the GAO SBR at the beginning of the cycle contained 850 mg/L of NaAc·3H$_2$O (12.5 C-mmol/L, approximately 400 mg COD/L) and 107 mg/L of NH$_4$Cl (2 N-mmol, 28 mg NH$_4^+$-N/L) as carbon and nitrogen sources, respectively. In order to suppress the growth of PAO and favor the development of an enriched GAO culture, phosphorus concentration in GAO SBR influent was limited to 2.2 mg PO$_4^{3-}$-P/L (0.07 P-mmol/L) leading to a P/COD influent ratio of about 1/200 (weight/weight) (Liu *et al.*, 1997). 2 mg/L of allyl-N-thiourea (ATU) were added to inhibit nitrification. The rest of minerals and trace metals present in the synthetic media were prepared as described by Smolders *et al.*, (1994). Prior to use, synthetic media were autoclaved at 110 °C for 1 h.

3.2.4. Analyses

The performance of the GAO SBR was regularly monitored by measuring orthophosphate (PO$_4^{3-}$-P), mixed liquor suspended solids (MLSS) and mixed liquor volatile suspended solids (MLVSS). In batch experiments, orthophosphate (PO$_4^{3-}$P), oxygen uptake rate (OUR), MLSS, MLVSS, poly-hydroxy-butyrate (PHB), poly-hydroxy-valerate (PHV), glycogen and ammonium (NH$_4^+$-N) concentrations were measured. All off-line analyses of orthophosphate, MLSS, MLVSS and ammonium were determined in accordance with Standard Methods (A.P.H.A., 1995). The OUR profiles were calculated based on the oxygen depletion (respirometric) profiles measured in the 10 mL-respirometer using linear regression and taking into account that the solubility of oxygen in water varies with temperature. The

PHA content (as PHB and PHV) of the freeze-dried biomass was determined according to the method described by Smolders *et al.* (1994). Glycogen was also determined following the method described by Smolders *et al.* (1994) but extending the digestion phase at 100°C from 1 to 5 h.

Fluorescence *in situ* Hybridization (FISH) was performed as described in Amman (1995). In order to determine the microbial population distribution in the GAO SBR, FISH probes used in this study were EUBMIX (mixture of probes EUB 338, EUB338-II and EUB338-III) to target the entire bacterial population (Amman *et al.*, 1995; Daims *et al.*, 1999); PAOMIX (mixture of probes PAO462, PAO651 and PAO 846) to target the *Betaproteobacteria Accumulibacter spp.* (Crocetti *et al.*, 2000); GAOMIX (equal amounts of GAOQ989 and GB_G2) for the *Gammaproteobacteria Competibacter spp.* (Crocetti *et al.*, 2002; Kong *et al.*, 2002); DF1MIX (TFO_DF218 plus TFO_DF218) for the *Alphaproteobacteria* from Cluster 1 *Defluviicoccus spp.* (Wong *et al.*, 2004); DF2MIX (DF988, DF1020 together with helper probes H966 and H1038) for Cluster 2 *Defluviicoccus spp.* (Meyer *et al.*, 2006); and, SBR9-1a for the *Alphaproteobacteria Sphingomonas spp.* (Beer *et al.*, 2004). Large aggregates were avoided by mild sonication (5W, 30 s). The quantification of the population distribution was carried out using the MATLAB image processing toolbox (The Mathworks, Natick, MA) as described in Lopez-Vazquez *et al.* (2007). Around fifteen separate randomly chosen images were evaluated with final results reflecting the average fractions of *Accumulibacter*, *Competibacter*, *Defluviicoccus* Cluster 1 and 2 and *Sphingomonas* present in corresponding activated sludge samples. Microbial population fractions were expressed as percentage of EUB. The standard error of the mean (SEM) was calculated as the standard deviation of the area percentages divided by the square root of the number of images analysed.

3.2.5. Aerobic batch test experiments

One hour before the end of the anaerobic phase, 450 mL of the GAO culture enriched at 20 °C in the parent SBR were manually transferred to the batch reactor. At this time, all HAc supplied at the beginning of the cycle had been totally taken up. The working temperature of the batch reactor (10, 20, 25, 30, 35 and 40 °C) was set 1 h before the sludge transfer. Once the sludge was transferred to the batch reactor, the sludge was stirred and kept under anaerobic conditions for 1 - 1.5 h before starting the corresponding aerobic batch test in order to adjust the microorganisms to the new temperature. During this period, N_2 gas was continuously sparged to keep anaerobic conditions and to avoid oxygen intrusion. After this acclimation period, the sludge was exposed to aerobic conditions. Air was continuously introduced into the reactor at a flow rate of 60 L/h and the sludge was constantly stirred at 500 rpm during the whole batch test. All aerobic batch tests experiments had a 2 h length. Once the aerobic conditions started, an extensive characterization of both liquid phase (orthophosphate, ammonium and OUR) and biomass (PHB, PHV, glycogen, MLSS and MLVSS) was performed throughout the 2-h aerobic batch tests.

3.2.6. Aerobic ATP maintenance coefficient of GAO

This experiment followed the experiment on aerobic stoichiometry and kinetics. As the sludge was already exposed to the aerobic environment for 2 h, the experiment continued by extended aeration for 24 h without substrate addition. A steady respiration rate in the absence of substrate indicated the oxygen requirements for cell maintenance or m_{OS} (the specific oxygen demand for maintenance, in mol O2/C-mol biomass/h). The energy required to cover the aerobic maintenance needs, as adenosine triphosphate (ATP), was expressed in terms of the aerobic maintenance coefficient m_{ATP}^{O} (in mol ATP/C-mol biomass/h units). For each aerobic batch test, the m_{ATP}^{O} coefficient was calculated based on the observed m_{OS} and according to the metabolic model of GAO (Zeng *et al.*, 2003).

3.2.7. Aerobic stoichiometry and kinetics

Glycogen production (r_{GLY}), OUR (r_{O2}), and PHA (r_{PHA}), PHB (r_{PHB}) and PHV (r_{PHV}) consumption, biomass production (r_X) and the oxygen consumed (O$_2$) per active biomass ratio (O$_2$/active biomass) were the kinetic rates of interest in this study. The maximum specific kinetic rates at the different studied temperatures were determined by linear regression based on the carbon profiles measured from the corresponding batch experiments. Maximum specific kinetic rates were expressed in C-mol/C-mol active biomass/h units. The active biomass was determined as the volatile suspended solids concentration excluding the PHB, PHV and glycogen contents (active biomass = MLVSS – PHB – PHV – glycogen). In order to calculate the specific rates, active biomass concentrations were expressed in C-mol units by taking into account the GAO biomass composition (CH$_{1.84}$O$_{0.5}$N$_{0.19}$) determined by Zeng *et al.* (2003).

Following the approach suggested by Zeng *et al.* (2003), biomass production (r_X) from each batch experiment was calculated based on the aerobic ammonium consumption measured during the aerobic tests considering that nitrification was inhibited due to the continuous addition of ATU in the influent. The O$_2$/active biomass ratio (in mol O$_2$/C-mol biomass units) was obtained by quantifying the total amount of oxygen consumed in each aerobic batch test (area below the OUR profile curve) excluding the oxygen requirements for maintenance.

The aerobic stochiometry was calculated based on the measured aerobic kinetic rates and using the overall equations for the conversion of PHA (r_{PHA}) and oxygen (r_{O2}) (Zeng *et al.*, 2003):

$$-r_{PHA} = \frac{1}{Y_{sgly}^{max}} r_{GLY} + \frac{1}{Y_{sx}^{max}} r_X + m_S \cdot X_{GAO} \tag{3.1}$$

$$-r_{O2} = \frac{1}{Y_{ogly}^{max}} r_{GLY} + \frac{1}{Y_{ox}^{max}} r_X + m_{OS} \cdot X_{GAO} \tag{3.2}$$

Where, r_{PHA} is the PHA degradation profile observed in the aerobic batch test ($r_{PHA} = r_{PHB} + r_{PHV}$); r_{O2} is the OURs profile computed during the 2-h aerobic test; Y_{sgly}^{max} and Y_{sx}^{max} are the maximum yields of glycogen and biomass production on PHA, respectively; m_S is the specific PHA demand for maintenance; Y_{ogly}^{max} and Y_{ox}^{max} are the maximum yields of glycogen and biomass production on oxygen, respectively; m_{OS} is the specific oxygen demand for maintenance; and, X_{GAO} is the active biomass concentration. Theoretically, m_S is the energy consumption for maintenance of the cell integrity (caused by basic metabolic energy requirements such as membrane potential, renewal of proteins, etc.) that, under the absence of external substrates, utilizes the internal stored substrate PHA (van Loosdrecht and Henze, 1999). Meanwhile, m_{OS} represents the oxygen demand of the cells for maintenance independently on whether external (HAc) or internal (PHA) substrates are present. All maximum conversion yields are expressed in C-mol/C-mol units. m_S and m_{OS} are in C-mol PHA/C-mol biomass/h and mol O_2/C-mol biomass/h units, respectively.

According to the aerobic metabolic models of PAO and GAO (Smolders *et al.*, 1994; Murnleitner *et al.*, 1997; Zeng *et al.*, 2003), all maximum aerobic yields are coupled, and function of the ATP/NADH ratio (δ). δ is a measure of the efficiency of the oxidative phosphorylation with a maximal theoretical value of 3. It indicates the amount of energy, as ATP, produced per nicotinamide adenine dinucleotide (NADH) oxidized in the oxidative phosphorylation. At pH = 7.0 and for an acetate-enriched GAO culture, all maximum yields can be expressed as function of δ as (Zeng *et al.*, 2003):

$$\frac{1}{Y_{sgly}^{max}} = \frac{2\delta + 1.26}{2.29\delta + 0.53} \tag{3.3}$$

$$\frac{1}{Y_{sx}^{max}} = \frac{2.13\delta + 2.29}{2.29\delta + 0.53} \tag{3.4}$$

$$m_S = \frac{m_{ATP}^O}{2.29\delta + 0.53} \tag{3.5}$$

$$\frac{1}{Y_{ogly}^{max}} = \frac{0.92}{2.29\delta + 0.53} \tag{3.6}$$

$$\frac{1}{Y_{ox}^{max}} = \frac{2.057}{2.29\delta + 0.53} \tag{3.7}$$

$$m_{OS} = \frac{1.146 m_{ATP}^O}{2.29\delta + 0.53} \tag{3.8}$$

If Equations 3.3 – 3.8 are substituted in Equations 3.1 – 3.2 the system is simplified to 2 equations and 6 variables (r_{PHA}, r_{O2}, r_{GLY}, r_X, δ and m_{ATP}^O) all of them, except δ, can be directly determined from the aerobic batch tests. This may simplify the system to 2 equations and 1 variable (δ). However, since fluctuations were observed in the aerobic glycogen profiles r_{GLY} was also considered as a variable. This led to a system of 2 equations and 2 variables (δ and r_{GLY}). AQUASIM (Reichert, 1994) was used to compute δ and r_{GLY} by simultaneously fitting the experimental time variable PHA concentrations and OUR observed in the aerobic batch tests to the r_{PHA} and r_{O2} profiles defined by the 2-equation system. The different kinetic expressions used to calculate δ and r_{GLY} are shown in Appendix 3.1.

3.2.8. Temperature coefficients

In this study, the effect of temperature on a constant rate relative to a standard temperature (here 20 °C) was expressed by the simplified Arrhenius equation:

$$r_T = r_{20} \cdot \theta_1^{(T-20)} \tag{3.9}$$

Where the r_T is the reaction at the temperature T, T is the temperature in °C and θ_1 is the temperature coefficient. Equation 3.9 proved to be suitable for fitting the results and allowed to comparing them with the temperature coefficients of different processes considered in mathematical models (Henze *et al.*, 2000).

In order to describe the kinetics of GAO for the whole experimental temperature range, a double Arrhenius expression was used:

$$r_T = r_{20} \cdot \theta_1^{(T-20)} \left[1 - \theta_2^{(T-T_{MAX})} \right] \tag{3.10}$$

In Equation 3.10, r_T is the reaction at the temperature T, T is the temperature in $^{\circ}C$, θ_1 is the temperature coefficient θ calculated from Equation 3.9, T_{MAX} is the temperature at which the microbial activity ceases, and θ_2 is a second temperature coefficient used to describe the declination in activity of the microorganisms at a temperature higher than the optimal (Bazin and Prosser, 2000).

3.3. Results

3.3.1. Enrichment of the GAO culture

The SBR was operated for more than 100 days before the aerobic batch tests were performed. It reached steady-state conditions within the first 50 days of operation. The biomass activity observed under steady-state conditions in the SBR displayed the characteristic phenotype of an enriched GAO culture (Figure 3.1a): complete anaerobic HAc uptake associated with glycogen consumption and PHA production (as PHB and PHV) and an anaerobic P-release less than 2 mg/L. Aerobically, the previously stored PHA was oxidized, glycogen was produced and ammonium was consumed. Nitrate and nitrite were not detected during the operation of the SBR. Average MLSS and MLVSS concentrations at the end of the aerobic phase were 2834 and 2653 mg/L, respectively. The MLVSS/MLSS ratio of 0.94 implied that a very low inorganic fraction (e.g. poly-phosphate) was stored. A quantification of the biomass population distribution obtained by FISH analysis indicated that *Competibacter* comprised around 93 ± 1 % of total bacterial population (as EUB) whereas *Accumulibacter* and *Defluviicoccus* were present in relatively low fractions (< 1 %) and *Sphingomonas* were not detected (Figure 3.1b). Other microorganisms, likely ordinary heterotrophs growing on lysis products, made up the rest of the microbial community (5 ± 1 %). Based on the biomass activity, low ash content and FISH results, it can be concluded that GAO were the dominant microorganisms present in the parent SBR.

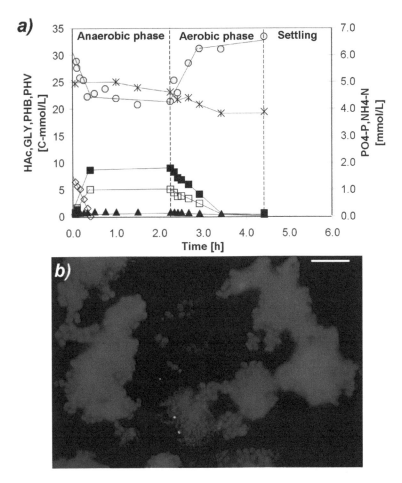

Figure 3.1. Glycogen Accumulating Organisms enriched in the parent sequencing batch reactor at 20 °C, pH 7.0 ± 0.1 and 10 d SRT: **(a)** cycle profiles observed in the enriched culture under steady state conditions: acetate (◊), glycogen (○), PHB (■), PHV (□) orthophosphate (▲) and ammonium (Ж); and **(b)** bacterial populations distribution by applying Fluorescence *in situ* Hybridization (bar indicates 10 μm, EUBacteria: blue, *Competibacter*: red, and *Accumulibacter* green. *Competibacter* appear violet due to superposition of EUBmix and GAOmix probes.

3.3.2. Aerobic stoichiometry of GAO

The temperature effects on the aerobic stoichiometry of GAO were evaluated considering that all aerobic metabolic processes are coupled and a function of the ATP/NADH ratio (δ) (Zeng *et al.*, 2003). Substituting Equations 3.3 to 3.8 in Equations 3.1 to 3.2, and employing the kinetic rates directly quantified from the results of the aerobic batch tests (r_{PHA}, r_X and m_{ATP}^{O}), led to a system of two equations with two unknown variables (δ and r_{GLY}). Using the parameter estimation toolbox of Aquasim (Reichert *et al.*, 1994) the 2-equation system was solved for each corresponding temperature by simultaneously fitting the experimental PHA and OUR profiles to the profiles calculated with the modified Equations 3.1 and 3.2. Figure 3.2 shows the experimental and fitted PHA, glycogen and OUR profiles from the aerobic batch test executed at 20 °C.

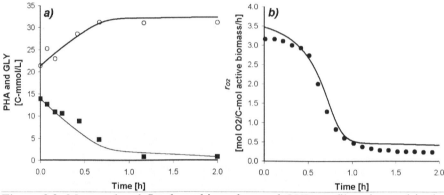

Figure 3.2. Measured and fitted aerobic carbon and OUR profiles observed in the aerobic batch test executed at 20 °C: (a) PHA (■) and glycogen (○) profiles; (b) measured OUR (●).

In average, the COD balances from the different aerobic batch tests closed to 97 ± 2 %. In order to verify the stoichiometry of the aerobic metabolic processes used in the current study, the different aerobic and anaerobic processes were integrated in a metabolic model for GAO (Lopez-Vazquez *et al.*, in preparation). At steady-state conditions (reached around 30 days), the GAO integrated metabolic model predicted an active biomass concentration of 87 C-mmol/L which is close to the average concentration of 96 ± 4 C-mmol/L observed in the GAO SBR (approximately 9 % difference).

The different estimated δ-values are shown in Figure 3.3. An average δ-value of 1.73 ± 0.37 was observed from 10 to 40 °C. This ratio is identical to the δ-value of 1.73 applied by Zeng *et al.* (2003) and in the range of the δ experimentally calculated by Smolders *et al.* (1994) (δ = 1.85) for PAO. Nevertheless, two distinctive trends were identified: from 10 to 30 °C an average δ of 1.95 ± 0.19 was measured while above 30 °C the average δ–value was 1.28 ± 0.03.

The maximum aerobic yields for each evaluated temperature were calculated using the different δ-values previously estimated and Equations 3.3, 3.4, 3.6

and 3.7. Table 3.1 presents the different maximum aerobic yields computed for the temperature interval of study. Regardless the different δ ratios observed from 10 to 40 °C, the aerobic stoichiometry seemed to be insensitive to temperature fluctuations when comparing the average yields from this study with the stoichiometric yields found by Zeng *et al.* (2003). Indeed, Y_{sgly}^{max} and Y_{sx}^{max} were insensitive to temperature. However, temperature had a slight influence on Y_{ogly}^{max} and Y_{ox}^{max} showing a decrease of about 30 % in the aerobic yields at temperatures higher than 30 °C compared to the trend observed below 30 °C (Table 3.1).

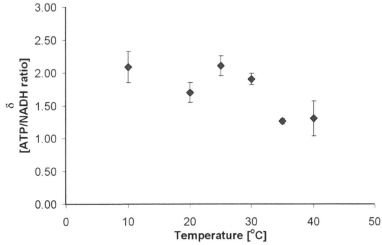

Figure 3.3. δ-ratios as function of temperature. Error bars indicate the standard deviation of the estimations.

Table 3.1. Aerobic stoichiometric parameters observed in this study.

Temperature [°C]	δ [a]	Y_{sgly}^{max} [b]	Y_{sx}^{max} [c]	Y_{ogly}^{max} [d]	Y_{ox}^{max} [e]	Reference
10	2.09	0.98	0.79	5.78	2.72	
20	1.70	0.96	0.75	4.96	2.27	
25	2.10	0.98	0.79	5.88	2.74	
30	1.90	0.98	0.78	5.57	2.53	This study
35	1.26	0.92	0.69	3.86	1.74	
40	1.30	0.92	0.70	3.92	1.78	
Average	1.73 ± 0.37	0.95 ± 0.03	0.75 ± 0.04	4.99 ± 0.91	2.30 ± 0.45	
20	1.73	0.95	0.75	4.89	2.18	Zeng *et al.* (2003)

[a] ATP/NADH ratio; [b] In: C-mol GLY/C-mol PHA units; [c] In: C-mol X/C-mol PHA units; [d] In: C-mol GLY/mol O_2 units; and, [e] In: C-mol X/ mol O_2 units.

3.3.3. Aerobic kinetics of GAO

Figure 3.4 displays the aerobic kinetic parameters measured in the different aerobic batch tests. In general, r_{GLY}, r_{PHA}, r_{PHB} and r_{PHV} (Figures 3.4a – 3.4d) increased gradually as temperature rose from 10 to 30 °C finding their optimal temperature at around 30 °C. Above 30 °C, bacterial activity declined but, even at 40 °C, it did not cease. As Table 3.2 shows, the r_{GLY} and r_{PHA} observed in this study at 20 °C are similar to the kinetic rates reported by Filipe *et al.* (2001b) but differ from the values presented by Zeng *et al.* (2003). Interestingly, the r_{PHA} reported by Smolders *et al.* (1995b) and Brdjanovic *et al.* (1997), which corresponds to an enriched PAO culture, resembles the r_{PHA} found in this paper at 20 °C; however, the r_{GLY} for PAO is lower likely due to the lower glycogen requirements of these organisms.

On the other hand, biomass production rates appeared to be insensitive to temperature from 20 to 35 °C meanwhile, at 10 and 40 °C, no biomass production was detected (Figure 3.4e). Unlike the other kinetic parameters, the O_2/biomass ratio increased continuously up to 40 °C without exhibiting any decline in activity (Figure 3.4f).

Table 3.2. Maximum aerobic kinetic rates at 20 °C and pH 7.0 observed in the present study and in other reports.

Organism	r_{pha} [a]	r_{gly} [b]	Reference
GAO	0.08	0.05	Zeng *et al.* (2003)[1]
	0.14 ± 0.01	0.14 ± 0.04	Filipe *et al.* (2001b)
	0.12 ± 0.01	0.13 ± 0.02	This study
PAO	0.09	-	Brdjanovic *et al.* (1997)
	0.12	0.04	Smolders *et al.* (1995b)

[1]. Kinetic rates calculated based on figures and data provided
[a] r_{pha} in C-mol PHA/C-mol X-h units; [b] r_{gly} in C-mol GLY/C-mol X-h units.

Excluding the O_2/biomass ratio, whose temperature coefficient was 1.046, the temperature coefficients up to 30 °C for the different kinetic parameters fluctuated between 1.077 and 1.090. Accordingly, the aerobic processes involved in the metabolism of GAO appear to have a medium degree of dependency on temperature (Henze *et al.*, 2000).

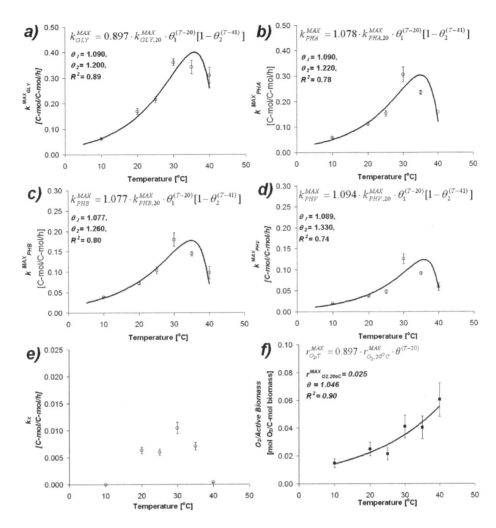

Figure 3.4. Effects of temperature on the aerobic kinetic rates of Glycogen Accumulating Organisms: **(a)** r_{GLY}, glycogen formation rate; **(b)** r_{PHA}, PHA formation rate; **(c)** r_{PHB}, PHB formation rate; **(d)** r_{PHV}, PHV formation rate; **(e)** r_X, biomass production rate; and, **(f)** O_2/active biomass ratio.

3.3.4. Aerobic ATP maintenance coefficient

m_{ATP}^O was moderately affected by temperature ($\theta = 1.081$) (Figure 3.5). In particular, the m_{ATP}^O observed at 20 °C (0.013 mol ATP/C-mol biomass/h) was in the range of values reported elsewhere for PAO and GAO enriched cultures (0.012 – 0.019 mol ATP/C-mol biomass/h) (Smolders *et al.*, 1995b; Brdjanovic *et al.*, 1997; Zeng *et al.*, 2003).

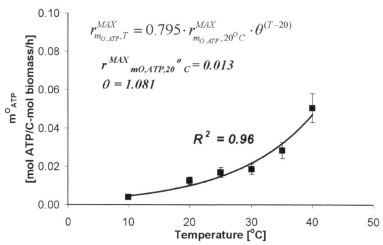

Figure 3.5. Temperature effects on the aerobic maintenance coefficient (m_{ATP}^{O}) of Glycogen Accumulating Organisms.

3.4. Discussion

3.4.1. Temperature effects on the aerobic stoichiometry

A series of aerobic batch tests from 10 to 40 °C were executed to address the temperature effects on the aerobic stoichiometry and kinetics of GAO. The short-term effects were studied in order to be able to evaluate the temperature effect on the metabolic rates without an influence of bacterial population changes that might occur with long-term studies. The average δ-value observed from 10 to 30 °C (1.95 ± 0.19) was in the range of different ratios found elsewhere (Table 3.3). However, the δ-values calculated at temperatures higher than 30 °C (1.28 ± 0.03) clearly differed from the reported ratios. A lower δ ratio suggests that the oxidative phosphorylation was less efficient as temperature increased over 30 °C. This was reflected in lower aerobic maximum yields (Table 3.1) and resulted in higher PHA and oxygen requirements for glycogen production, biomass growth and maintenance. In addition, as temperature raised above 30 °C a decline in r_{PHA} and r_{GLY} was observed (Figure 3.4) as well as a continuous increase on m_{ATP}^{O} (Figure 3.5). This is in accordance with Tijhuis *et al.* (1993) who concluded that temperature has a major influence on the maintenance requirements of microorganisms at temperatures higher than 30 °C, affecting the biomass growth rates. Correspondingly, Krishna and van Loosdrecht (1999a, 1999b) observed lower polymer transformation rates and higher maintenance requirements as temperature rose from 15 to 35 °C leading to lower biomass growth yields.

Table 3.3. Comparison between different δ-ratios reported in the literature.

System	δ [a]	Calculation method	Reference
Aerobic activated sludge	1.60	Estimated based on biomass activity and metabolic pathways	Beun *et al.* (2000)
Aerobic activated sludge	2.56 ± 0.08	Estimated based on biomass activity	Sin *et al.* (2005)
Pure culture (*Paracoccus pantotrophs*)	1.84	Estimated based on biomass activity and metabolic pathways	van Aalst-van Leewen *et al.* (1997)
Aerobic activated sludge	1.40-1.80	Estimated based on biomass activity and metabolic pathways	Dircks *et al.* (2001)
Enriched PAO culture	1.85	Directly quantified	Smolders et. al (1994)
Enriched PAO culture	1.46	Directly quantified	Brdjanovic *et al.* (1997)
Enriched GAO culture	1.74	Estimated based on biomass activity and metabolic pathways	Filipe *et al.* (2001d)
Enriched GAO culture	1.73	Average δ-ratio from Beun *et al.* (2000), Smolders *et al.* (1994) and Dircks. *et al.* (1997)	Zeng *et al.* (2003)
Enriched GAO culture	1.73 ± 0.37	Estimated based on biomass activity and metabolic pathways	Average ratio observed in this study

[a] ATP/NADH ratio.

On the basis of the short-term temperature experiments executed in the present study, we conclude that the aerobic stoichiometry of GAO is insensitive to temperature changes from 10 to 30 °C whereas temperature causes a major effect on the aerobic stoichiometric parameters of GAO as it increases above 30 °C. Long-term cultivation studies need to be carried out to assess whether a GAO population can adapt to these temperatures.

3.4.2. Temperature effects on the aerobic kinetics

Observed r_{GLY} as well as r_{PHA} were different from the rates reported by Zeng *et al.* (2003) (Table 3.2). On the basis of this study a direct explanation could not be found. It could possibly be related to the PHA content of the cells (f^{PHA}). For PAO cultures, f^{PHA}, which is dependent on the substrate feed to the system and biomass content, determines the reaction rates of the aerobic processes (Smolders *et al.*, 1995a). Zeng *et al.* (2003) applied an aerobic HRT of 12 h (6 cycles of 4 h per day with a 2 h aerobic phase per cycle) while in this study, likewise Filipe *et al.* (2001b), an aerobic HRT of 9 h was used (4 cycles of 6 h per day with a 2.25 h aerobic stage per cycle). In the three lab set-ups, the initial acetate concentration per cycle was similar (6.25 C-mmol/L) and, in combination with the active biomass concentration and aerobic HRT, resulted in lower initial aerobic f^{PHA} and PHA load in the system of Zeng and colleagues (2003) (0.12 C-mol PHA/C-mol X and 1.4 C-mol PHA/C-mol X/d, respectively) compared to the fractions and loads observed in the current study (0.17 C-mol PHA/C-mol X, 1.8 C-mol PHA/C-mol X/d) and elsewhere (0.20 C-mol PHA/C-mol X, 2.1 C-mol PHA/C-mol X/d) (Filipe *et al.*, 2001b). This might suggest that aerobic reaction rates proceed slower as result of lower initial aerobic f^{PHA} and PHA load.

Despite that the polymer transformation rates suffered a decline in activity at temperatures higher than 30 °C, temperature seemed not to have caused any effect on the biomass production rates between 20 and 35 °C (Figure 3.4e). There is a possibility that a consistent trend was not found due to small differences in measured values that may lay within the margin of error of the analytical technique (Brdjanovic *et al.*, 1997). Nevertheless, between 20 and 30 °C, the biomass growth rates were similar and had a relatively low standard deviation (± 15 %) which suggests that the errors in the analytical determination technique were not very high.

At 10 and 40 °C, no biomass growth was detected (Figure 3.4e). Either GAO's growth was too low, being difficult to measure, or temperature had an inhibitory effect restricting their growth at these temperatures. A severe increase in maintenance requirements, as previously discussed, seemed to be the reason of the limited growth of GAO at 40 °C. Interestingly, in a previous report (Lopez-Vazquez *et al.*, 2007), 40 °C had a detrimental effect on the anaerobic metabolism of GAO. However, in the present study, aerobic activity was still observed at this temperature (Figure 3.4) although, as previously mentioned, no biomass growth was observed.

The temperature coefficients found for the different GAO aerobic processes (θ = 1.046 − 1.082) are in the range of the coefficients reported by Brdjanovic *et al.* (1997) for PAO (θ = 1.035 − 1.081) indicating that PAO and GAO have similar aerobic temperature dependencies from 10 to 30 °C. This implies that the temperature effects on the anaerobic metabolism of PAO and GAO (Lopez-Vazquez *et al.*, 2007) have a higher influence on the PAO-GAO competition than the aerobic temperature effects.

3.4.3. Implications on the PAO-GAO competition

The results of our study suggest that the anaerobic and aerobic metabolisms of GAO have different temperature dependencies. This may have important implications on the PAO-GAO competition influencing the stability of the EBPR process.

Regarding GAO's anaerobic metabolism, GAO have important metabolic advantages over PAO from 20 to 35 °C (Lopez-Vazquez *et al.*, 2007). Below 20 °C, both microorganisms exhibit similar temperature dependencies concerning the maximum HAc uptake rate; however, PAO seemed to be favored since they present lower anaerobic maintenance requirements (Lopez-Vazquez *et al.*, 2007) and GAO exhibit a low biomass growth rate (this study). Therefore, taking into account the combined temperature effects on the anaerobic and aerobic metabolisms, temperatures higher than 20 °C are more favorable for GAO showing an optimal temperature around 30 °C, whereas, below 20 °C, PAO have metabolic advantages with an optimal temperature that seems to lay around 20 °C (Lopez-Vazquez *et al.*, 2007; Brdjanovic *et al.*, 1997). These findings could help to explain the stability of the EBPR process in colder regions (air temperature below 20 °C) (van Veldhuizen *et al.*, 1999; Ybstebφ *et al.*, 2000; Meijer *et al.*, 2002; Tykesson *et al.*, 2005).

Concerning the temperature effects in warm regions (temperature higher than 20 °C), Panswad *et al.* (2003), in a long-term study, observed that PAO were outcompeted by GAO in an EBPR SBR operated at 30 °C and 10 d SRT. However, the GAO culture could not be sustained when switching the temperature from 30 to 35 °C. The dominance and disappearance of GAO could be explained on the basis of the temperature dependencies observed in the present study: (a) the optimal temperature for GAO around 30 °C, and (b) the lower aerobic maximum yields and higher maintenance requirements as temperature rises above 30 °C.

In a similar study, Whang and Park (2006) observed a population shift from an enriched-GAO to an enriched-PAO culture at 30 °C when decreasing the SRT from 10 to 3 d at pH 7.5. Whang *et al.* (2007) hypothesized that a lower GAO biomass yield on PHA than that of PAO might cause the change in population when the SRT was shortened. However, according to the present study, the optimal temperature of GAO seems to be found around 30 °C, a reduction in the maximum aerobic yields was only observed above 30 °C, and biomass growth rates appear to be independent of temperature changes from 20 to 35 °C. At the applied operating conditions (3 d SRT and pH 7.5), the observations of Whang and Park (2006) imply that GAO have a higher minimum SRT than PAO. Likely, other factors besides temperature, such as the higher pH (7.5) applied in the study of Whang and Park (2006), might have also played an important role favoring the metabolism of PAO as described by Filipe *et al.* (2001c).

A long-term (weeks) study should be beneficial not only to help to elucidate the temperature effects on actual biomass production rates as consequence of the thermal effects on the different aerobic metabolic processes (among other unexplained observations that remained in this research), but also to confirm the temperature dependencies of GAO. Moreover, in a long-term study, potential population changes and adaptations that cannot be accounted for with short-term studies may be expected.

The present study was carried out using an enriched GAO culture where *Candidatus Competibacter Phosphatis* were the dominant microorganisms (93 ± 1 % with respect to total bacterial population). Therefore, on the basis of the current study it can not be concluded whether or not other groups of GAO would display a similar temperature dependence.

3.5. Conclusions

The aerobic stoichiometry of GAO was insensitive to temperature changes between 10 and 30 °C whereas it was affected at 35 and 40 °C. An overall optimal temperature for both anaerobic and aerobic metabolism of GAO appears to be found around 30 °C. The kinetic processes involved in the aerobic metabolism of GAO had a medium degree of dependency on temperature from 10 to 30 °C ($\theta = 1.046 - 1.082$) comparable to those of PAO. Within a broad temperature interval that covers the operating range of most of domestic WWTP (from 10 to 30 °C, and perhaps even at higher temperatures), we conclude that GAO do not have aerobic metabolic advantages over PAO, and that competitive advantages are due to anaerobic processes.

Acknowledgements

The authors acknowledge the National Council for Science and Technology (CONACYT, Mexico) and the Autonomous University of the State of Mexico for the scholarship awarded to Carlos Manuel Lopez Vazquez. Special thanks to the lab-staff from UNESCO-IHE Institute for Water Education.

References

APHA/AWWA (1995) Standard methods for the examination of water and wastewater, 19th ed. Port City Press, Baltimore.

Amann R. I. (1995). *In situ* identification of microorganisms by whole cell hybridization with rRNA-targeted nucleic acid probes. In: Akkermans ADL, van Elsas JD, de Bruijn FJ, editors. Molecular Microbial Ecology Manual. London: Kluwer Academic Publisher. p 1-15.

Bazin MJ, Prosser JI (2000) Physiological models in microbiology. CRC Series in mathematical models in microbiology. Florida: CRC Press.

Beer M, Kong, YH, Seviour RJ (2004) Are some putative glycogen accumulating organisms (GAO) in anaerobic: aerobic activated sludge systems members of the *α-Proteobacteria*? Microbiology-SGM 150:2267–2275.

Beun JJ, Paletta F, van Loosdrecht MCM, Heijnen JJ (2000) Stoichiometry and kinetics of poly-β-hydroxybutyrate metabolism in aerobic, slow growing, activated sludge cultures. Biotechnol Bioeng 67(4):379-388.

Brdjanovic D, van Loosdrecht MCM, Hooijmans CM, Alaerts GJ, Heijnen JJ (1997) Temperature effects on physiology of biological phosphorus removal. ASCE J Environ Eng 123(2):144-154.

Brdjanovic D, Logemann S, van Loosdrecht MCM, Hooijmans CM, Alaerts GJ, Heijnen JJ (1998a) Influence of temperature on biological phosphorus removal: process and molecular ecological studies. Water Res 32(4):1035-1048.

Cech JS, Hartman P (1993) Competition between phosphate and polysaccharide accumulating bacteria in enhanced biological phosphorus removal systems. Water Res 27:1219-1225.

Crocetti GR, Banfield JF, Keller J, Bond PL, Blackall LL (2002) Glycogen accumulating organisms in laboratory-scale and full-scale wastewater treatment processes. Microbiol 148:3353-3364.

Crocetti GR, Hugenholtz P, Bond PL, Schuler A, Keller J, Jenkins D, Blackall LL (2000) Identification of polyphosphate-accumulating organisms and design of 16S rRNA-directed probes for their detection and quantitation. Appl Environ Microbiol 66(3):1175–1182.

Daims H, Bruhl A, Amman R, Schleifer KH, Wagner M (1999) The domain-specific probe EUB 338 is insufficient for the detection of all bacteria: development and evaluation of a more comprehensive probe set. Syst Appl Microbiol 22:434-444.

Dircks K, Beun JJ, van Loosdrecht MCM, Heinen JJ, Henze M (2001) Glycogen metabolism in aerobic mixed cultures. Biotechnol Bioeng 73(2):85-94.

Erdal UG, Erdal ZK, Randall CW (2003) The competition between PAO (phosphorus accumulating organisms) and GAO (glycogen accumulating organisms) in EBPR (enhanced biological phosphorus removal) systems at different temperatures and the effects on system performance. Water Sci Technol 47(11):1-8.

Filipe CDM, Daigger GT, Grady Jr CPL (2001a) A metabolic model for acetate uptake under anaerobic conditions by glycogen-accumulating organisms: stoichiometry, kinetics and effect of pH. Biotechnol Bioeng 76(1):17-31.

Filipe CDM, Daigger GT, Grady Jr CPL (2001b) Effects of pH on the aerobic metabolism of phosphate-accumulating organisms and glycogen-accumulating organisms. Water Environ Res 73(2):213-222.

Filipe CDM, Daigger GT, Grady Jr CPL (2001c) pH as a key factor in the competition between glycogen-accumulating organisms and phosphorus-accumulating organisms. Water Environ Res 73(2):223-232.

Filipe CDM, Daiggert GT, Grady CPL (2001d) An integrated metabolic model for the anaerobic and aerobic metabolism of glycogen accumulating organisms. In: 3rd International Water Association specialized conference on microbiology of the activated sludge and biofilm processes. Rome, Italy.

Henze M, Gujer W, Mino T, van Loosdrecht MCM (2000) Activated sludge models ASM1, ASM2, ASM2d and ASM3. IWA Scientific and Technical Report No. 9.

IWA Task Group on Mathematical Modelling for Design and Operation of Biological Wastewater Treatment. London: IWA Publishing.

Kong YH, Ong SL, Ng WJ, Liu WT (2002) Diversity and distribution of a deeply branched novel proteobacterial group found in anaerobic–aerobic activated sludge processes. Environ Microbiol 4(11): 753–757.

Krishna C, van Loosdrecht MCM (1999a) Effect of temperature on storage polymers and settleability of activated sludge. Water Res 33(10):2374-2382.

Krishna C, van Loosdrecht MCM (1999b) Substrate flux into storage and growth in relation to activated sludge modeling. Water Res 33(14):3149-3161.

Lopez-Vazquez CM, Song YI, Hooijmans CM, Brdjanovic D, Moussa MS, Gijzen HJ, van Loosdrecht MCM (2007) Short-term temperature effects on the anaerobic metabolism of glycogen accumulating organisms. Biotech Bioeng 97(3):483-495.

Liu WT, Nakamura K, Matsuo T, Mino T (1997) Internal energy-based competition between polyphosphate- and glycogen-accumulating bacteria in biological phosphorus removal reactors-effect of P/C feeding ratio. Water Res 31(6):1430-1438.

Meijer SCF, van Loosdrecht MCM, Heijnen JJ (2002) Modelling the start-up of a full-scale biological nitrogen and phosphorus removing WWTP. Water Res 36(19):4667-4682.

Meyer RL, Saunders AM, Blackall LL (2006) Putative glycogen accumulating organisms belonging to *Alphaproteobacteria* identified through rRNA-based stable isotope probing. Microbiology-SGM 152:419–429.

Mino T, van Loosdrecht MCM, Heijnen JJ (1998) Microbiology and biochemistry of the enhanced biological phosphorus removal process. Water Res 32(11):3193-3207.

Murnleitner E, Kuba T., van Loosdrecht MCM, Heijnen JJ (1997) An integrated metabolic model for the aerobic and denitrifying biological phosphorus removal. Biotechnol Bioeng 54(5):434-450.

Panswad T, Doungchai A, Anotai J (2003) Temperature effect on microbial community of enhanced biological phosphorus removal system. Water Res 37:409-415.

Reichert P (1994) AQUASIM - a tool for simulation and data analysis of aquatic systems. Water Sci Technol 30(2): 21-30.

Satoh H, Mino T, Matsuo T (1994) Deterioration of enhanced biological phosphorus removal by the domination of microorganisms without polyphosphate accumulation. Water Sci Technol 30(6):203-211.

Sin G, Guisasola A, De Pauw DJW, Baeza JA, Carrera J, Vanrolleghem PA (2005) A new approach for modelling simultaneous storage and growth processes for activated sludge systems under aerobic conditions. Biotechnol Bioeng 92(5):600-613.

Smolders GJF, van der Meij J, van Loosdrecht MCM, Heijnen JJ (1994) Stoichiometric model of the aerobic metabolism of the biological phosphorus removal process. Biotechnol Bioeng 44:837-848.

Smolders GJF, Klop JM, van Loosdrecht MCM, Heijnen JJ (1995a) Validation of the metabolic model: effect of the sludge retention time. Biotechnol Bioeng 48(3):222-233.

Smolders GJF, van der Meij J, van Loosdrecht MCM, Heijnen JJ (1995b) A structured metabolic model for anaerobic and aerobic stoichiometry and kinetics of the biological phosphorus removal process. Biotechnol Bioeng 47(3):277-287.

Tijhuis J, van Loosdrecht MCM, Heijnen JJ (1993) A thermodinamically based correlation for maintenance Gibbs energy requirements in aerobic and anaerobic chemotrophic growth. Biotechnol Bioeng 42(4):509-519.

Tykesson E, Jönsson L-E, la Cour Jansen J (2005) Experience from 10 years of full-scale operation with enhanced biological phosphorus removal. Water Sci Technol 52(12):151-159.

van Aalst-van Leewen MA, Pot MA, van Loosdrecht MCM, Heijnen JJ (1997) Kinetic modeling of poly (β-hydroxybutyrate) production and consumption by *Paracoccus pantotrophus* under dynamic substrate supply. Biotechnol Bioeng 55(5):773-782.

van Loosdrecht MCM and Henze M (1999) Maintenance, endogenous respiration, lysis, decay and predation. Wat Sci Technol 39(1):107-117.

van Veldhuizen HM, van Loosdrecht MCM, Heijnen JJ (1999) Modelling biological phosphorus and nitrogen removal in a full scale activated sludge process. Water Res 33(16):3459-3468.

Whang LM, Filipe CDM, Park JK (2007) Model-based evaluation of competition between polyphosphate- and glycogen-accumulating organisms. Water Res 41(6):1312-1324.

Whang LM, Park JK (2006) Competition between polyphosphate- and glycogen-accumulating organisms in enhanced biological phosphorus removal systems: effect of temperature and sludge age. Water Environ Res 78(1):4-11.

Wong MT, Tan FM, Ng WJ, Liu WT (2004) Identification and occurrence of tetrad-forming alphaproteobacteria in anaerobic–aerobic activated sludge processes. Microbiology-SGM 150:3741–3748.

Ydstebɸ L, Bilstad T, Barnard J (2000) Experience with biological nutrient removal at low temperatures. Water Environ Res 72(4):444-454.

Zeng RJ, van Loosdrecht MCM, Yuan Z, Keller J (2003) Metabolic model for glycogen-accumulating organisms in anaerobic/aerobic activated sludge systems. Biotechnol Bioeng 81(1):92-105.

Appendix 3.1.

Stoichiometric matrix and kinetic expressions used in the present study.

Process	Component				Kinetic expression
	X_{PHA}	X_{GLY}	X_{GAO}	S_{O2}	
Glycogen production	$-\dfrac{1}{Y_{SGLY}^{MAX}}$	1		$-\dfrac{1}{Y_{OGLY}^{MAX}}$	$r_{GLY}\dfrac{(f_{GLY}^{MAX}-f_{GLY})}{(f_{GLY}^{MAX}-f_{GLY})+K_{GLY}}\dfrac{X_{PHA}}{K_{PHA}+X_{PHA}}X_{GAO}$
Biomass growth	$-\dfrac{1}{Y_{SX}^{MAX}}$		1	$-\dfrac{1}{Y_{OX}^{MAX}}$	$r_{X}\dfrac{X_{PHA}}{K_{PHA}+X_{PHA}}X_{GAO}$
Maintenance	$-m_S$			$-m_{OS}$	X_{GAO}

4

Chapter

Long-term temperature influence on the metabolism of GAO

Content

This chapter has been accepted for publication as:
    Lopez-Vazquez CM, Hooijmans CM, Brdjanovic D, Gijzen HJ, van Loosdrecht MCM (2008) Temperature effects on glycogen accumulating organisms. *Water Research* (in press). DOI: 10.1016/j.watres.2009.03.038.

Abstract

The long-term temperature effects on the anaerobic and aerobic stoichiometry and conversion rates on adapted enriched cultures of *Competibacter* (a known GAO) were evaluated from 10 to 40 °C. The anaerobic stoichiometry of *Competibacter* was constant from 15 to 35 °C, whereas the aerobic stoichiometry was insensitive to temperature changes from 10 to 30 °C. At 10 °C, likely due the inhibition of the anaerobic conversions of *Competibacter*, a switch in the dominant bacterial population to an enriched *Accumulibacter* culture (a known PAO) was observed. At higher temperatures (35 and 40 °C), the aerobic processes limited the growth of *Competibacter*. Due to the inhibition or different steady-state (equilibrium) conditions reached at long-term by the metabolic conversions, the short- and long-term temperature dependencies of the anaerobic acetate uptake rate of *Competibacter* differed considerably between each other. Temperature coefficients for the various metabolic processes are derived, which can be used in activated sludge modeling. Like for PAO cultures: (i) the GAO metabolism appears oriented at restoring storage pools rather than fast microbial growth, and (ii) the aerobic growth rate of GAO seems to be a result of the difference between PHA consumption and PHA utilization for glycogen synthesis and maintenance. It appears that the proliferation of *Competibacter* in EBPR systems could be suppressed by adjusting the aerobic solids retention time while, aiming at obtaining highly enriched PAO cultures, EBPR lab-scale reactors could be operated at low temperature (e.g. 10 °C).

4.1. Introduction

Different operating and environmental conditions have been identified as determinant factors to understand the competition between PAO and GAO (Oehmen *et al.*, 2007). Among them, temperature appears to play an important role on the interaction between these two types of microorganisms (Panswad *et al.*, 2003; Erdal *et al.*, 2003; Whang and Park, 2006; Lopez-Vazquez *et al.*, 2007, 2008). All these studies agree that PAO have important advantages over GAO at low and moderate temperatures (below 20 °C), while higher temperatures (higher than 20 °C) are more beneficial for GAO. However, those studies were carried out either without using a highly enriched GAO culture or for short-term periods. Thus, although relevant conclusions regarding the temperature effects on GAO have been obtained, there is a clear need to study the long-term (weeks) temperature effects on their metabolisms in order to evaluate possible biomass adaptations and bacterial population changes that only can be observed at long-term. Moreover, the execution of long-term experiments could help to address and elucidate unexplained observations that remained from previous studies (Panswad *et al.*, 2003; Whang and Park, 2006; Lopez-Vazquez *et al.*, 2007, 2008).

In the current chapter, the long-term (weeks) temperature effects on (1) the anaerobic and (2) aerobic metabolisms of GAO, regarding their stoichiometry and kinetics, were evaluated. The research was carried out using a lab-enriched GAO culture within a temperature interval (10 to 40 °C) that covers the operating temperature of most of the wastewater treatment plants. The temperature coefficients obtained in this study can be incorporated into activated sludge models to describe the activity of GAO. This could help to improve our understanding about the occurrence of GAO at different temperatures aiming at the optimization of the EBPR process operation under different weather conditions.

4.2. Materials and methods

4.2.1. Continuous operation of the parent sequencing batch reactor (SBR1)

A GAO culture was enriched in a double-jacketed lab-scale sequencing batch reactor (SBR). This SBR, hereafter referred to as SBR1 or parent reactor, had a working volume of 2.5 L. The SBR1 was operated at 20 ± 0.5 °C and pH 7.0 ± 0.1 in cycles of 6 hours (2.25 h anaerobic, 2.25 h aerobic and 1.5 h settling phase). At the beginning of each cycle, 1.25 L of synthetic medium was fed to the parent reactor over a period of 5 minutes. 312.5 mL of mixed liquor were removed on a daily basis (about 78 mL per cycle) from SBR1, resulting in a solids retention time (SRT) of approximately 8 days. Further details regarding the operation of SBR1 can be found in Chapter 2. The long-term temperature effect tests were performed in a separate SBR (SBR2). Thus, SBR1 was continuously operated during the whole experiment and only used as source of enriched GAO biomass.

4.2.2. Operation of the parallel sequencing batch reactor (SBR2)

In order to evaluate the long-term temperature effects on the anaerobic and aerobic metabolism of GAO, a separate SBR (SBR2) was used. SBR2 was physically identical to SBR1 and also operated under similar conditions: a 6 hour-cycle (2.25 h anaerobic, 2.25 h aerobic and 1.5 h settling phase) and at a pH 7.0 ± 0.1. For the execution of the long-term experiments, the temperature of SBR2 was set according to the desired temperature of study (10, 15, 30, 35 or 40 °C). In order to avoid temperature fluctuations, prior to filling up the reactor, the temperature of the synthetic influent was adjusted to that of study in a third double-jacketed batch reactor equipped with a water bath.

4.2.3. Long-term temperature effect tests

The long-term temperature effect tests were carried out in SBR2 using the GAO culture cultivated in the parent reactor (SBR1). Figure 4.1 displays a schematic diagram of the execution of the five long-term temperature effect tests performed in the current study (at 10, 15, 30, 35 and 40 °C). As displayed in Figure 4.1, first, the enriched GAO culture was cultivated in SBR1. Once the GAO biomass reached steady-state conditions, it was transferred from SBR1 to SBR2 and acclimatized for at least 8 days at 20 °C using a SRT of 8 days. Meanwhile, after the sludge transfer, SBR1 was reinoculated with mixed liquor that had been previously removed (waste of sludge) to control the SRT. After the biomass activity of SBR2 reached steady-state conditions, the temperature was switched to any of the evaluated temperatures (10, 15, 30, 35 and 40 °C). Measurements of the first cycles were carried out immediately after the temperature was switched aiming at determining the temperature effects on the enriched biomass in the first cycle and comparing them with previous short-term studies (Lopez-Vazquez *et al.*, 2007, 2008). The initial SRT of the SBR2 was set at 8 days. Each long-term test was carried out for a time period equivalent to 3 times the SRT, or longer, until the biomass activity reached steady-state conditions. During the execution of certain long-term tests, the SRT had to be increased in order to keep the GAO enriched culture. In those tests where the SRT was modified, the operation of SBR2 was extended for an equivalent period of 3 times the adjusted SRT, or until reaching steady-state conditions. After concluding with the evaluation of any of the studied temperatures, the SBR2 was thoroughly cleaned before executing other temperature effect test.

4.2.3. Synthetic media

In order to suppress the growth of PAO, phosphorus concentration in the synthetic media supplied to both SBR1 and SBR2 was limited to 2.2 mg PO_4^{3-}-P/L (0.07 P-mmol/L) (Liu *et al.*, 1997). The synthetic media also contained per litre: 850 mg $NaAc \cdot 3H_2O$ (12.5 C-mmol, approximately 400 mg COD/L) and 107 mg NH_4Cl (2 N-mmol). 0.002 g of allyl-N-thiourea (ATU) were added to inhibit nitrification. The rest of minerals and trace metals present in the synthetic media were prepared as described by Smolders *et al.*, (1994). Prior to use, synthetic media were autoclaved at 110 °C for 1 h.

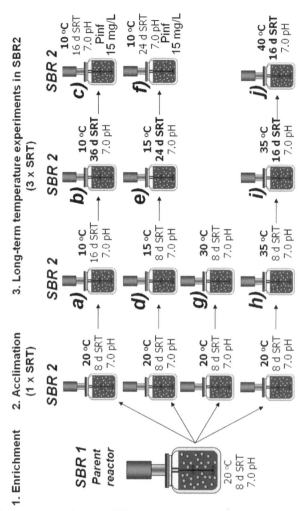

Figure 4.1. Experimental design followed in the present study for the execution of the long-term temperature tests.

4.2.4. Analyses

The performance of SBR1 and SBR2 was regularly monitored through the determination of orthophosphate (PO_4^{3-}-P), mixed liquor suspended solids (MLSS) and mixed liquor volatile suspended solids (MLVSS). During cycle measurements, orthophosphate (PO_4^{3-}P), acetate (HAc), MLSS, MLVSS, poly-hydroxy-butyrate (PHB), poly-hydroxy-valerate (PHV), glycogen and ammonium (NH_4^+-N) concentrations as well as oxygen uptake rates (OUR) were measured. All off-line analyses were performed in accordance with Standard Methods (A.P.H.A., 1995). The OUR, and PHA (as PHB and

PHV) and glycogen contents were determined as described elsewhere (Lopez-Vazquez *et al.*, 2007, 2008).

In order to determine the microbial population distributions in the different long-term tests, Fluorescence *in situ* Hybridization (FISH) was performed as described in Amman (1995). FISH probes used in this study were EUBMIX (mixture of probes EUB 338, EUB338-II and EUB338-III) to target the entire bacterial population (Daims *et al.*, 1999); PAOMIX (mixture of probes PAO462, PAO651 and PAO 846) to target the *Betaproteobacteria Accumulibacter spp.* (Crocetti *et al.*, 2000); GAOMIX (equal amounts of GAOQ989 and GB_G2) for the *Gammaproteobacteria Competibacter spp.* (Crocetti *et al.*, 2002; Kong *et al.*, 2002); DF1MIX (TFO_DF218 plus TFO_DF218) for the *Alphaproteobacteria* from Cluster 1 *Defluviicoccus spp.* (Wong *et al.*, 2004); and, DF2MIX (DF988, DF1020 together with helper probes H966 and H1038) for Cluster 2 *Defluviicoccus spp.* (Meyer *et al.*, 2006). Large aggregates were avoided by mild sonication (5W, 30 s). The quantification of the population distribution was carried out as described in Lopez-Vazquez *et al.* (2008).

4.2.5. Anaerobic stoichiometry and kinetics

Net P-released, glycogen hydrolysis and PHB and PHV production per HAc consumed (P/HAc, glycogen/HAc, PHB/HAc, PHV/HAc ratios, respectively) were the anaerobic stoichiometric parameters of interest in this study. These parameters were calculated taking into account the initial HAc concentration and biomass composition (as MLSS, MLVSS, glycogen, PHB and PHV) at the beginning and end of the anaerobic phase.

The maximum acetate consumption rate ($q_{SA,GAO}^{MAX}$) at the different studied temperatures was determined considering the HAc consumption profile and the biomass concentration at the beginning of the corresponding experiment. Like all kinetic rates in the present study, $q_{SA,GAO}^{MAX}$ was expressed in C-mol/(C-mol /h) units, as described elsewhere (Lopez-Vazquez *et al.*, 2007).

4.2.6. Aerobic stoichiometry and kinetics

Glycogen production (r_{GLY}), OUR (r_{O2}), the consumption rates of PHA (r_{PHA}), PHB (r_{PHB}) and PHV (r_{PHV}), and the biomass production (r_X) were the aerobic rates of interest. The maximum specific rates at the different studied temperatures were determined by linear regression based on the carbon profiles observed at the corresponding cycle measurements and expressed in C-mol/C-mol/h units (Lopez-Vazquez *et al.*, 2008). r_X was calculated based on the aerobic ammonium consumption measured during the aerobic tests considering that nitrification was inhibited due to the continuous addition of ATU in the influent (Zeng *et al.*, 2003). The net biomass growth rate was also determined based on the average biomass concentration and considering the amount of mixed liquor daily wasted to control the SRT.

The aerobic stoichiometric ratios of interest were: the maximum yields of glycogen and biomass production on PHA (Y_{sgly}^{max} and Y_{sx}^{max}, respectively); and, the maximum yields of glycogen and biomass production on oxygen (Y_{ogly}^{max} and Y_{ox}^{max}, respectively). The specific PHA and oxygen demands for maintenance (m_S and m_{OS}, respectively) were not experimentally determined in the present study. Instead, the values calculated in a previous study were used for the calculation of the aerobic yields (Lopez-Vazquez *et al.*, 2008). All maximum conversion yields are expressed in C-mol/C-mol units, whereas m_S and m_{OS} in C-mol PHA/(C-mol biomass-h) and mol O_2/(C-mol biomass-h) units, respectively. According to the aerobic metabolic models of PAO and GAO (Smolders *et al.*, 1995; Murnleitner *et al.*, 1997; Zeng *et al.*, 2003), all maximum aerobic yields are coupled, and function of the ATP/NADH$_2$ ratio (δ), which is a measure of the efficiency of the oxidative phosphorylation. Following the same approach like in a previous study (Lopez-Vazquez *et al.*, 2008), the maximum aerobic stoichiometric yields of the different long-term temperature tests were calculated through the determination of the δ-ratio using AQUASIM (Reichert, 1994).

4.2.7. Temperature coefficients

The simplified Arrhenius expression was used to describe the effect of temperature on the conversion rates of biomass:

$$r_T = r_{20} \cdot \theta^{(T-20)} \tag{4.1}$$

In Equation 4.1, r_T is the reaction at the temperature T, T is the temperature in °C and θ is the temperature coefficient.

Whenever possible, the kinetics of GAO for the whole experimental temperature range (from 10 to 40 °C) was described using an extended Arrhenius equation (modified from Bazin and Prosser, 2000):

$$r_T = r_{20} \cdot \theta_1^{(T-20)} \left[1 - \theta_2^{(T-T_{MAX})} \right] \tag{4.2}$$

Where θ_1 is the temperature coefficient θ calculated from Equation 1, T_{MAX} is the temperature at which the microbial activity ceases, and θ_2 is a second temperature coefficient used to describe the decline in biomass activity at temperatures higher than the optimal.

4.3. Results

4.3.1. Enrichment of the GAO culture

The biomass activity from SBR1 (or parent reactor) was already under steady-state conditions for more than 100 days when the present study started (Lopez-Vazquez *et al.*, 2008). It showed the typical phenotype of an enriched GAO culture (Figure 4.2): complete HAc uptake associated with glycogen consumption, PHA production and a low anaerobic P-released/HAc uptake ratio in the anaerobic stage (0.03 P-mol/C-mol) and, under aerobic conditions, glycogen replenishment and PHA degradation. According to FISH analyses, *Competibacter* made up to 93 ± 1 % (relative to EUBMIX) and *Accumulibacter* about 3 ± 1 % of the total bacterial population in SBR1. *Defluvicoccus* were present in negligible fractions (< 1 %). Other microorganisms, likely ordinary heterotrophs, comprised the rest of the microbial community (3 ± 1 %). The quantification of the bacterial populations, together with the observed biomass activity, confirmed that SBR1 was highly enriched with GAO, being *Competibacter* the dominant microorganisms.

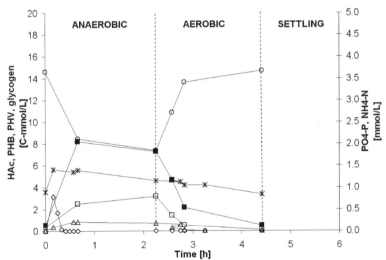

Figure 4.2. Cycle profiles observed in the enriched culture of glycogen accumulating organisms (*Competibacter*) cultivated in the parent sequencing batch reactor (SBR1) at 20 °C, pH 7.0 ± 0.1 and 8 days SRT under steady state conditions: acetate (◊), glycogen (○), PHB (■), PHV (□) orthophosphate (Δ) and ammonium (Ж).

As previously mentioned, the sludge cultivated in SBR1 was only used as a source of enriched GAO biomass to perform the long-term temperature effect tests in the parallel reactor (SBR2). Therefore, SBR1 was continuously operated during the whole experimental period (for more than 400 days) exhibiting a stable activity (data not shown).

4.3.2. Cultivation of the enriched GAO culture at different temperatures

As indicated in Figure 4.1, during the execution of the long-term temperature effect tests, the SRT of SBR2 were extended, due to the lower unexpected biomass growth rates, from the initially proposed SRT of 8 days to 16, 24 and even 36 days (Figures 4.1b, e and i). The SRT length was increased when the biomass was unable to take up the entire carbon source in the anaerobic stage for more than 12 cycles (72 h). However, as described later, the GAO enriched culture could not be cultivated at certain temperatures despite that a longer SRT was applied.

The biomass activities measured in the first cycle, just after switching the temperature from 20 °C to the studied temperatures, as well as the long-term biomass activities, are shown in Appendix A (Figures 4.Aa, d, f, h and j). *Competibacter* were satisfactorily cultivated at 15, 30 and 35 °C (Figures 4.Ae, g, and i), but not at 10 and 40 °C (Figures 4.Ab and j) since the metabolism of GAO was severely affected.

At 10 °C and despite that a relatively low influent P/HAc ratio was supplied (0.006 P-mol/C-mol), the inhibitory temperature effects on GAO resulted favorable for the proliferation of PAO (*Accumulibacter*). At 10 °C and 16 days SRT (Figure 4.1a), HAc started to leak into the aerobic stage just 5 days after the long-term experiment started. Thereupon, the sludge wastage was stopped in order to extend the SRT, which increased from 16 to approximately 36 days. 16 days later, HAc continued to leak into the aerobic phase, meanwhile the anaerobic P/HAc ratio rose from 0.05 P-mol/C-mol (first cycle, Figures 4.1a and Aa) to 0.13 P-mol/C-mol (Figures 4.1b and 4.Ab) indicating that PAO started to proliferate growing on the carbon source that was not consumed by GAO. Interestingly, also a considerable accumulation of glycogen and PHA was observed implying that these two polymers were not being consumed (Figure 4.Ab). The *Competibacter* and *Accumulibacter* fractions changed from 93 ± 1 % and 3 ± 1 %, respectively (estimated in the first cycle, Figure 4.3a), to 43 ± 2 % *Competibacter* and 8 ± 1 % *Accumulibacter* after 16 days (Figure 4.3b). The low EBPR biomass activity implied that PAO were being suppressed by the low influent P concentration (Figure 4.Ab). On day 16th, aiming at evaluating whether PAO could grow under these conditions and outcompete *Competibacter*, the influent P concentration was increased to 15 mg/L (0.48 P-mol) (Smolders *et al.*, 1994) and the SRT re-adjusted to 16 days (Figure 4.1c). On day 32nd, *Accumulibacter* comprised around 37 ± 2 % of total biomass, whereas *Competibacter* made up to 12 ± 1 % (Figure 4.3c). On day 64th, the biomass activity was already under steady-state conditions showing the typical phenotype of an enriched PAO culture (Figure 4.Ac): considerable anaerobic P/HAc ratio (0.56 P-mol/C-mol) as well as complete aerobic P-removal. FISH analysis confirmed that *Accumulibacter* were the dominant microorganisms (79 ± 1 %), meanwhile the *Competibacter* fraction was around 18 ± 1 % (Figure 4.3d).

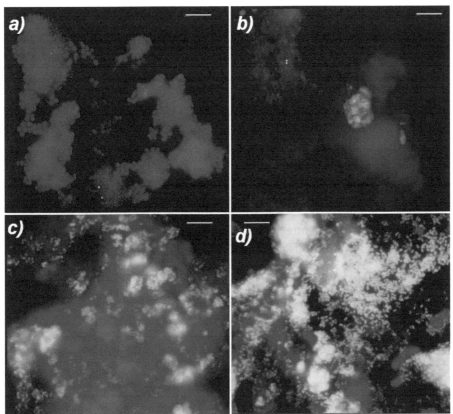

Figure 4.3. Bacterial population distributions observed in the long-term temperature effect test carried out at 10 °C by applying Fluorescence *in situ* Hybridization (bar indicates 10 μm, EUBacteria: blue, *Competibacter*: red, and *Accumulibacter:* green): (a) microbial populations at 20 °C under steady-state conditions before switching the temperature to 10 °C; (b) on day 16th after switching the temperature to 10 °C; (c) on day 32nd, and (d) on day 64th. *Competibacter* appears violet due to superposition of EUBmix and GAOmix probes, while *Accumulibacter* light green due to the superposition of EUBMIX and PAO mix probes.

After the completion of the long-term experiment at 15 °C (Figure 4.1e), the temperature-effect test on GAO at 10 °C was repeated (Figure 4.1f). The objective was to evaluate whether a GAO culture acclimatized at a lower temperature (in this case 15 °C) could be cultivated at 10 °C or PAO would become again the dominant microorganisms. Thereby, after studying the long-term temperature effects at 15 °C, the temperature was switched to 10 °C and the influent phosphorus concentration increased to 0.48 P-mol (15 mg PO_4^{3-}-P/L). The SRT was kept at 24 days. After 1 SRT (24 days) the anaerobic P/HAc ratio was 0.58 P-mol/C-mol and a complete aerobic P-removal was observed (data not shown) suggesting that PAO had become the dominant microorganisms. These results confirmed that *Accumulibacter* had considerable metabolic advantages over *Competibacter* at 10 °C.

At 40 °C (Figures 4.1j and 4.Aj), the activity of *Competibacter* could not be kept for more than 6 cycles (less than 2 days). During this period, the maximum acetate uptake rate rapidly decreased, resulting in acetate leak into the aerobic stage. After 2 days, the waste of sludge was stopped to extend the applied SRT; however, the biomass activity was not recovered. Due to the detrimental effect, the long-term temperature effects on *Competibacter* could not be studied at this relatively high temperature (40 °C).

Excluding the changes in populations observed at 10 °C, FISH analyses indicated that *Competibacter* were the dominant organisms in all long-term temperature studies. The average microbial populations observed at 15, 30 and 35 °C, after reaching steady-state conditions, were similar to the population distributions observed at 20 °C: 94 ± 1 % *Competibacter*, 2 ± 2 % *Accumulibacter* and the rest was comprised by other microorganisms (4 ± 2 %). *Defluviicoccus* were not observed at 10 °C and only a few cells were hardly detected at 15, 30 and 35 °C (comprising less than 1 % of the total bacterial population).

4.3.3. Long-term temperature effects on the anaerobic stoichiometry

The temperature effects on the anaerobic stoichiometry of *Competibacter* are displayed in Table 4.1. In average, all COD balances closed to 95 ± 3 %, suggesting that most of the intracellular compounds involved in the anaerobic metabolism of GAO were satisfactorily measured. Due to the switch in bacterial populations from an enriched GAO to an enriched PAO culture, the stoichiometric ratios from the cycles carried out at 10 °C were rather different compared to the rest of the temperatures, and even between each other (Table 4.1). At 10 °C, the anaerobic glycogen hydrolysis seemed to be inhibited as suggested by the lower measured anaerobic glycogen/HAc ratio (0.96 C-mol/C-mol). Due to unknown reasons, higher anaerobic ratios were measured in the long-term test carried out at 30 °C that considerably differed from the main trends. The long-term effects at 40 °C could not be evaluated, since, as previously described, this temperature caused a detrimental effect on *Competibacter*. Nevertheless, the anaerobic stoichiometric parameters from the first cycle were similar to those observed in other long-term studies (Table 4.1), implying that, likely, the anaerobic stochiometry of GAO was not inhibited and, therefore, might have not been the main cause of the deterioration of the activity of *Competibacter* at 40 °C.

In general, the average anaerobic transformations of intracellular stored polymers glycogen, PHA, PHB, PHV) per HAc consumed measured in the first and long-term cycles (Table 4.1) showed that the anaerobic stoichiometry of *Competibacter* is rather insensitive to temperature changes from 15 to 35 °C and comparable to those observed in previous studies (Zeng *et al.*, 2003; Lopez-Vazquez *et al.*, 2007).

4.3.4. Temperature effects on the maximum acetate uptake rate

The long-term temperature effects on the maximum acetate uptake rates of *Competibacter* ($q_{SA,GAO}^{MAX}$) were studied from 10 to 40 °C. Figure 4.4 shows the short-term temperature dependency of $q_{SA,GAO}^{MAX}$ as observed by Lopez-Vazquez *et al.* (2007) (Figure 4.4a), the $q_{SA,GAO}^{MAX}$ measured in the first cycles of the long-term tests (Figure 4.4b), and the uptake rates observed in the long-term studies after reaching steady-state conditions (Figure 4.4c). The temperature dependencies observed in the short-term and in the first cycles of the long-term tests were relatively similar (Figures 4.4a and 4.4b), as indicated by their comparable temperature coefficients (1.054 and 1.074, respectively). However, the temperature dependency of $q_{SA,GAO}^{MAX}$ in the long-term tests was considerably different (Figure 4.4c): (1) from 10 to 20 °C, $q_{SA,GAO}^{MAX}$ increased from a negligible uptake rate of 0.02 to 0.20 C-mol/C-mol/h, which is similar to that observed by Zeng *et al.* (2003) and Oehmen *et al.* (2005) at 20 °C; (2) from 20 and 35 °C, the uptake rates appeared to be relatively constant at about 0.20 C-mol/C-mol/h; and, (3) above 35 °C, $q_{SA,GAO}^{MAX}$ suddenly decreased, ceasing at around 40 °C. The differences in the temperature dependencies of $q_{SA,GAO}^{MAX}$ clearly indicate that the results from short- and long-term studies can lead to different conclusions as biomass adapts or acclimatizes to different temperatures.

Table 4.1. Anaerobic stoichiometric parameters observed in the present study and in previous reports.

Temperature [°C]	Cycle	Organisms	SRT [d]	gly/HAc	PHA/HAc	PHB/HAc	PHV/HAc	PHV/PHB	P/HAc	Reference
						[mol/mol]				
10	First	Competibacter	8	0.96	1.95	1.49	0.46	0.31	0.05	This study
	Intermediate	Competibacter and Accumulibacter	16	0.62	1.42	1.10	0.32	0.29	0.13	
	Long-term steady-state	Accumulibacter	16	0.55 ± 0.05	1.44 ± 0.07	1.31 ± 0.05	0.13 ± 0.02	0.10 ± 0.02	0.56 ± 0.02	
15	First	Competibacter	8	1.06	1.76	1.29	0.47	0.36	0.03	
	Long-term steady-state	Competibacter	24	1.23 ± 0.12	1.81 ± 0.14	1.29 ± 0.02	0.52 ± 0.02	0.40 ± 0.03	0.03 ± 0.00	
20	Parent SBR steady-state cycle	Competibacter	8	1.21 ± 0.15	1.88 ± 0.03	1.45 ± 0.02	0.44 ± 0.02	0.30 ± 0.01	0.03 ± 0.00	
30	First	Competibacter	8	1.15	2.05	1.54	0.50	0.33	0.00	
	Long-term steady-state	Competibacter	8	1.68 ± 0.22	2.33 ± 0.02	1.60 ± 0.05	0.74 ± 0.04	0.46 ± 0.05	0.01 ± 0.00	
35	First	Competibacter	8	1.36	2.13	1.38	0.75	0.55	0.01	
	Long-term steady-state	Competibacter	16	1.12 ± 0.04	1.91 ± 0.12	1.22 ± 0.01	0.68 ± 0.00	0.56 ± 0.00	0.01 ± 0.01	
40	First	Competibacter	16	1.32	2.12	1.44	0.67	0.47	0.00	
10 – 40	**Average First cycles**	**Competibacter**	-	**1.17 ± 0.17**	**2.00 ± 0.15**	**1.43 ± 0.10**	**0.57 ± 0.13**	**0.40 ± 0.10**	**0.02 ± 0.02**	
15 – 40*	**Average Long-term cycles**	**Competibacter**	-	**1.19 ± 0.06**	**1.87 ± 0.05**	**1.32 ± 0.12**	**0.55 ± 0.12**	**0.42 ± 0.13**	**0.02 ± 0.01**	
10 – 40	Short-term tests	Competibacter	8	1.20 ± 0.19	1.97 ± 0.13	1.28 ± 0.06	0.69 ± 0.07	0.54 ± 0.04	0.01	Lopez-Vazquez et al. (2007a)
20	Long-term	GAO	6.6	1.12	1.86	1.40	0.47	0.34	0.00	Zeng et al. (2003)
20	Long-term	PAO	8	0.50	1.33	1.33	0.00	0.00	0.50	Smolders et al. (1994a)

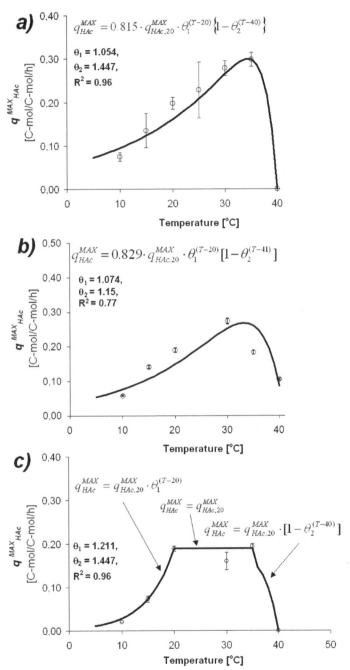

Figure 4.4. Temperature effects on the maximum acetate uptake rate of *Competibacter* ($q_{SA,GAO}^{MAX}$) observed in: (a) previous short-term temperature effect tests (Lopez-Vazquez *et al.*, 2007); (b) first cycles of the present study; and, (c) long-term, under steady-state conditions.

4.3.5. Long-term temperature effects on the aerobic stoichiometry

In the long-term temperature effect tests, at 10 °C, the aerobic stoichiometric coefficients were different because of the change in bacterial populations from a culture where *Competibacter* were dominant to an enriched *Accumulibacter* culture (Figure 4.3). As consequence, the aerobic stoichiometric parameters measured in the long-term study carried out at 10 °C (Table 4.2) were similar to those reported for PAO at 20 °C (Smolders *et al.*, 1995). From 15 to 30 °C, the aerobic stoichiometry of *Competibacter* appeared to be insensitive to temperature changes (Table 4.2). This is in accordance with previous observations (Lopez-Vazquez *et al.*, 2008). At 35 °C, the ATP/NADH$_2$ ratio (1.30 ± 0.04) determined at long-term was lower than that observed from 15 to 30 °C (1.73), implying that the oxidative phosphorylation was less efficient. This was reflected in the lower aerobic maximum yields. Since the GAO culture could not be cultivated at 40 °C, the long-term effects on the aerobic stoichiometry of GAO were not evaluated at this temperature. Nevertheless, the δ-value measured in the first cycle executed at 40 °C was lower (1.52 ± 0.12), which suggested that the relatively high temperature immediately affected the aerobic stoichiometry of *Competibacter*. The decreased energy yield might have limited the cultivation of *Competibacter* at 40 °C.

4.3.6. Long-term temperature effects on the aerobic kinetics

As the temperature coefficients indicate, the maximum PHA degradation rate (k_{PHA}^{MAX}) measured in the first cycles and long-term studies (Figures 4.5b and 4.5c) showed similar temperature dependencies than those reported for short-term experiments in a previous study (Figure 4.5a) (Lopez-Vazquez *et al.*, 2008). The temperature dependencies of the glycogen formation rates (k_{GLY}^{MAX}) were also comparable (Figures 4.5d, e and f). A combined Arrhenius expression (Equation 2) satisfactorily described the temperature dependencies of k_{PHA}^{MAX} and k_{GLY}^{MAX} observed in the long-term studies for the whole studied temperature range (from 10 to 40 °C).

Table 4.2. Aerobic stoichiometric parameters observed in the present study and in previous reports.

Temperature [°C]	Cycle	Organisms	P/O ratio (δ) [ATP/NADH$_2$]	Y_{SGLY}	Y_{SX}	Y_{SPP}	Y_{OGLY}	Y_{OX}	Y_{OPP}	Reference
				[mol/mol]						
10	First cycle	*Competibacter*	1,73	0,95	0,75	-	4,81	2,26	-	This study
	Long-term cycle	*Accumulibacter*	1,85	0,94	0,75	3,71	3,92	2,36	3,28	
15	First cycle	*Competibacter*	1,73	0,95	0,75	-	4,86	2,27	-	
	Long-term cycle	*Competibacter*	1,73	0,96	0,76	-	4,96	2,29	-	
20	Parent SBR steady-state cycle	*Competibacter*	1,73	0,95	0,75	-	4,89	2,28	-	
30	First cycle	*Competibacter*	1,73	0,96	0,76	-	5,01	2,30	-	
	Long-term cycle	*Competibacter*	1,73	0,96	0,76	-	4,98	2,29	-	
35	First cycle	*Competibacter*	1,73	0,96	0,76	-	5,03	2,30	-	
	Long-term cycle	*Competibacter*	1,30 ± 0,04	0,92	0,70	-	3,93	1,78	-	
40	First cycle	*Competibacter*	1,52 ± 0,12	0,94	0,73	-	4,43	2,04	-	
10 – 40	Short-term tests	*Competibacter*	1,73 ± 0,37	0,95	0,75	-	4,89	2,18	-	Lopez-Vazquez et al. (2008)
20	Long-term	PAO	1,85	0,90	0,74	3,68	3,92	2,44	3,27	Smolders et al. (1994b)
20	Long-term	GAO	1,73	0,95	0,75	-	4,89	2,18	-	Zeng et al. (2003)

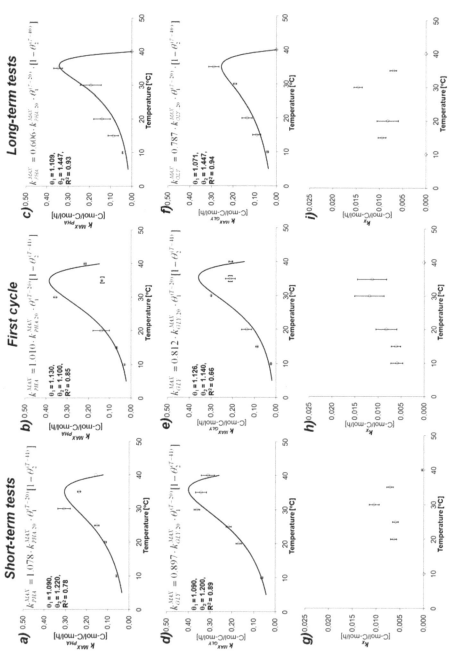

Figure 4.5. Temperature effects on the maximum aerobic PHA degradation (k_{PHA}^{MAX}: a-c), aerobic glycogen production (k_{GLY}^{MAX}: d-f) and biomass production rates (k_X: g-i) of *Competibacter* observed in previous short-term temperature effect tests (Lopez-Vazquez et al., 2008), and in the first and long-term cycles of the present study.

The maximum growth rates (k_X) calculated in the different experiments (short-term, first cycles and long-term; Figures 4.5g, 4.5h and 4.5i, respectively) did not show any consistent trend that may help to explain why different SRT had to be applied in the long-term tests. Moreover, the k_X observed at 15, 20 and 35 °C were relatively similar (Figure 4.5i), indicating that the same SRT could have been used for the cultivation of GAO at these temperatures whereas, actually, longer SRT had to be applied at 15 and 35 °C (24 and 16 d, correspondingly) compared to that at 20 °C (8 d). Murnleitner *et al* (1997) already indicated that the growth rate of PAO is a resultant of the PHA degradation, and poly-P and glycogen formation. Therefore, the minimum SRT for PAO had to be calculated from a model (Brdjanovic *et al.*, 1998b). Likely, the same applies for the case of GAO.

4.4. Discussion

4.4.1. Anaerobic stochiometry of GAO

The temperature effects on the metabolism of GAO are rather complex not only because most of the involved metabolic reactions take place on intracellular stored polymers, such as PHA and glycogen, but also due to the different temperature dependencies of the anaerobic and aerobic processes.

The lower glycogen/HAc ratio measured in the first cycle at 10 °C (0.96 C-mol/C-mol, Table 4.1) indicated that the glycolysis metabolic pathway was partially inhibited. This was also observed in a previous study (Lopez-Vazquez *et al.*, 2007). The inhibition of glycolysis could have affected the metabolism of GAO, reducing the production of the required energy and electrons for substrate uptake under anaerobic conditions. This might have resulted in the lower observed maximal acetate uptake rate (Figure 4.4). The accumulation of glycogen observed in the intermediate cycle (Figure 4.Ab), which increased up to 55 C-mol/L, appeared to confirm the inhibition of the glycolysis pathway. After the influent phosphorus concentration was increased (Figures 4.1c and 4.Ac), PAO (*Accumulibacter*) started to proliferate in the culture. This indicates that PAO are more able to adapt and grow at low temperatures than GAO. Likely, aiming at obtaining highly enriched PAO cultures, EBPR lab-scale reactors could be operated at low temperature (e.g. 10 °C) as a strategy to suppress the growth of GAO.

On the basis of the long-term temperature effect tests carried out in the present study, it can be concluded that the anaerobic stoichiometry of *Competibacter* is insensitive to temperature changes from 15 to 35 °C.

4.4.2. Maximum acetate uptake rate

In the long-term experiments, the temperature dependency of the maximum acetate uptake rate ($q_{SA,GAO}^{MAX}$), particularly at temperatures higher than 20 °C (Figure 4.4c), was rather different than the temperature dependencies observed in previous short-term tests and in the first cycles of the long-term tests carried out in the present research (Figures 4.4a and 4.4b). As discussed above, the lower acetate uptake rate measured in the long-term tests at 10 °C could be a consequence of the inhibition of the glycolysis metabolic process. The relatively constant acetate uptake rate (at about 0.19 C-mol/C-mol/h) observed from 20 to 35 °C can not be explained on the basis of this study. Taking into account that glycogen is assumed to be utilized as energy and carbon source for both anaerobic substrate uptake and anaerobic maintenance (Filipe *et al.*, 2001; Zeng *et al.*, 2002), likely, an increase in the anaerobic maintenance requirements limited the substrate uptake process since, as temperature rises, the anaerobic maintenance requirements also increase and more glycogen should be consumed to cover the anaerobic maintenance needs leaving less glycogen available for substrate uptake (Lopez-Vazquez *et al.*, 2007). At 40 °C, the effect of higher maintenance requirements could be even more notorious because the $q_{SA,GAO}^{MAX}$ observed in the first cycle (Figure 4.Aj) rapidly decreased during the first eight cycles of the long-term experiment, leading to HAc leak into the aerobic stage and complete deterioration of the GAO activity (data not shown). Overall, it seems that the glycogen conversion rate is the rate limiting process in the anaerobic metabolism of GAO.

4.4.3. Aerobic stoichiometry of GAO

In order to address the long-term temperature effects on the aerobic stoichiometry and kinetics of GAO, experiments were executed from 10 to 40 °C. Despite that *Competibacter* could not be cultivated at 10 °C, the aerobic parameters measured in the first cycle at 10 °C (Table 4.2) were similar to those observed by Zeng *et al.* (2003). This suggests that the relatively low temperature (10 °C) apparently did not affect the aerobic stoichiometry of *Competibacter* and, therefore, this seemed not to have been the cause that led to the dominance of *Accumulibacter*. From 15 to 30 °C, the aerobic stoichiometry of GAO was not influenced by temperature changes (Table 4.2). At 35 °C, the ATP/NADH$_2$ ratio (δ-values) calculated from the long-term study (1.30 ± 0.04) was lower, resulting in lower aerobic maximum yields, and higher PHA and oxygen requirements for glycogen production and biomass growth (Table 4.2). Correspondingly, Krishna and van Loosdrecht (1999a, 1999b) and Lopez-Vazquez *et al.* (2008) observed lower polymer transformation rates and higher maintenance requirements as temperature rose above 30 °C, leading to lower biomass growth yields. In a similar study (Panswad *et al.* 2003), a GAO culture could not be sustained when switching the temperature from 30 to 35 °C using a SRT of 10 days. In the current long-term study at 35 °C, the SRT needed to be extended from 8 to 16 days in order to keep the enriched *Competibacter* culture. This corroborates that, likely, *Competibacter* had a lower net biomass growth at 35 °C due to an increase in the aerobic maintenance requirements and less efficient energy generation.

We conclude that the aerobic stoichiometry of GAO is insensitive to temperature changes from 15 to 30 °C, and likely even from 10 °C, whereas temperature causes a major effect on the aerobic stoichiometry of GAO as it increases above 30 °C.

4.4.4. Aerobic kinetics of GAO

Based on the temperature coefficients, the long-term temperature experiments indicated that k_{PHA}^{MAX} and k_{GLY}^{MAX} (Figures 4.5c and 4.5f) have a medium to high degree of dependency on temperature which is comparable to that proposed for ordinary heterotroph organisms, fermentation and nitrification processes (1.07 – 1.12; Henze *et al.*, 2000). The results of the present study show that temperature effects on the metabolism of GAO are rather complex because the different aerobic metabolic processes have different temperature dependencies. This should be taken into account when studying or modeling the temperature effects on the PAO-GAO competition and, thus, the occurrence of *Competibacter* in EBPR systems.

4.4.5. Temperature effects on the PAO-GAO competition

As observed by Zeng *et al.* (2003), the PAO-GAO competition can take place (i) in the anaerobic stage, through the competition for substrate, and, (ii) in the aerobic phase, considering that the biomass production takes place under aerobic conditions. Thus, the organism that can take up substrate faster in the anaerobic stage and/or have a higher biomass production in the aerobic phase may have important advantages to lead the competition.

In Figure 4.6, a comparison between the maximum HAc uptake rates of PAO and GAO ($q_{SA,PAO}^{MAX}$ and $q_{SA,GAO}^{MAX}$, respectively) observed in previous (Brdjanovic *et al.* 1997, 1998a; Lopez-Vazquez *et al.*, 2007) and in the present short- and long-term studies are displayed. Contrary to the uptake rates of GAO measured at short- (black line) and long-term (black bold line), the rates of PAO determined from short- and long-term studies were similar (gray bold line). Concerning the maximum substrate uptake rates, *Competibacter* has important kinetic advantages over PAO at short-term, as temperature raises over 20 °C. At long-term, PAO had a higher substrate uptake at temperatures below 20 °C. Above 20 °C, although the HAc uptake rate of GAO measured at long-term noticeably decreased compared to that observed at short-term, it is still around 12 % higher than that of PAO (0.19 and 0.17 C-mol/C-mol/h, respectively), resulting in a kinetic advantage for *Competibacter* at temperatures higher than 20 °C. Although the maximum acetate uptake rate of GAO tends to be similar from 20 to 35 °C, it can not be discarded that a higher rate might exist around 25 °C as the long-term effect of this temperature has not been studied on these organisms.

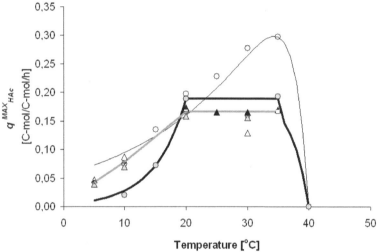

Figure 4.6. Temperature effects on the maximum acetate uptake rates of PAO ($q_{SA,PAO}^{MAX}$) and GAO ($q_{SA,GAO}^{MAX}$): (open triangles) short-term effects on PAO, Brdjanovic et al. (1997); (gray closed triangles) long-term temperature effects on PAO, Brdjanovic et al., 1998; (black closed triangles) short-term effects on PAO, Lopez-Vazquez et al. (2007); (open circles) short-term effects on *Competibacter*; (gray closed circles) long-term effects on *Competibacter*, this study. Gray bold line: trend line from short- and long-term effects of PAO. Black thin line: trend line of the short-term effects of *Competibacter*; and, black bold line: trend line of the long-term temperature effects on *Competibacter*.

The net biomass production at different temperatures can be assumed to be a function of (i) the temperature dependency of the biomass production rate and (ii) the PHA available for biomass growth, which depends upon the temperature effects on the aerobic metabolic processes (e.g. PHA degradation, glycogen formation and aerobic maintenance requirements). Thus, the net biomass production rate appears to be a resultant of the combination of the temperature effects on the aerobic metabolic processes of

Competibacter. Brdjanovic *et al.* (1998b) hypothesized that, for enriched PAO cultures, the minimum aerobic SRT determines the presence of PAO in a system since biomass growth takes place under aerobic conditions and depends on the degradation rate and availability of PHA. Therefore, Brdjanovic and colleagues (1998b) proposed an expression to determine the minimum required aerobic SRT (SRT_{MIN}^{AER}) of PAO at different temperatures in terms of (i) the net biomass production observed per cycle per HAc consumed under steady-state conditions ($Y_{X.HAc}$); (ii) the maximum PHA degradation rate (k_{PHA}^{MAX}); and, (iii) the PHA availability (accounted for through the maximum PHA present in the cells, f_{PHA}^{MAX} , and the anaerobic PHA production per HAc consumed, $Y_{PHA.HAc}$). Further details can be found elsewhere (Brdjanovic *et al.*, 1998b). Following a similar approach (Appendix B), the minimum aerobic SRT of *Competibacter* ($SRT_{MIN,GAO}^{AER}$) was determined from 10 to 40 °C and compared to that of PAO ($SRT_{MIN,PAO}^{AER}$) (Figure 4.7). Since PAO have not been cultivated above 30 °C following a systematic study (Brdjanovic *et al.*, 1998a, 1998b), $SRT_{MIN,PAO}^{AER}$ at temperatures higher than 30 °C could not be computed. The calculated minimal SRT for *Competibacter* fitted satisfactorily between the different successful (closed squares) and unsuccessful attempts (open squares) to cultivate a GAO culture (Figure 4.7).

In general, PAO require a shorter minimum aerobic SRT than *Competibacter* (Figure 4.7). The difference is larger at temperatures lower than 20 °C. This could be explained based on the aerobic kinetics of *Competibacter* (Figure 4.5c and 4.5f): k_{PHA} and k_{GLY} are so similar below 20 °C that most of the PHA is utilized for glycogen replenishment rather than for biomass production or maintenance, limiting the growth of these microorganisms (Figure 4.7). The glycogen accumulation observed in the long-term studies at 10 and 15 °C (Figures 4.Ab and 4.Ae, respectively) confirms that the aerobic glycogen formation was preferred over biomass production. Interestingly, Murnleitner *et al.* (1997) observed a similar trend on enriched PAO cultures.

From 20 to 30 °C, the growth of PAO is still larger than that of *Competibacter* (Figure 4.7). This suggests that, in lab- and full-scale EBPR systems, *Competibacter* could be outcompeted by applying an aerobic SRT between $SRT_{MIN,PAO}^{AER}$ and $SRT_{MIN,GAO}^{AER}$. Accordingly, Whang and Park (2006) observed a switch in the dominant microbial population from an enriched-GAO to an enriched-PAO culture at 30 °C after shortening the SRT from 10 to 3 days, resulting in a nominal aerobic SRT of 1.8 days. Whang *et al.*

(2007) hypothesized that, although GAO had a higher substrate uptake rate, the lower biomass growth of GAO was the main cause that led to the dominance of PAO. Potentially, short aerobic SRT could be applied in lab-scale systems to favor the development of highly enriched PAO cultures. However, the difference between the minimum aerobic SRT of PAO (1.25 days) and GAO (1.9 days) is likely too small to be effectively used in practical situations (e.g. full-scale EBPR activated sludge systems). Furthermore, in full-scale systems performing nitrification, the minimum SRT of nitrifying bacteria will be likely longer.

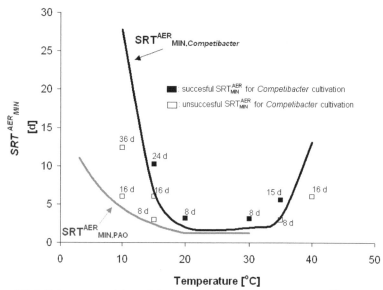

Figure 4.7. Minimum aerobic solid retention times of PAO ($SRT_{MIN,PAO}^{AER}$; gray line) and GAO ($SRT_{MIN,GAO}^{AER}$; black line) as a function of the temperature, as calculated by Brdjnanovic et al. (1998b) and in the current study, respectively. Closed squares (■) indicate the successful (long enough) and open squares (□) the unsuccessful (too short) aerobic SRT applied for the cultivation of *Competibacter*. Closed and open squares refer to the aerobic SRT (y-axis), while numbers next to the squares indicate the total applied SRT in the experiments.

As previously discussed, due to an increase in the aerobic maintenance requirements of *Competibacter* (Krishna and van Loosdrecht, 1999a, 1999b; Lopez-Vazquez *et al.*, 2008) there appeared to be less PHA available for biomass growth at temperatures higher than 30 °C, resulting in a lower biomass production and, as consequence, in a longer $SRT_{MIN,GAO}^{AER}$ as temperature raised over 30 °C (Figure 4.7).

Considering the lower maximum substrate uptake rate and longer minimum aerobic SRT of *Competibacter* observed at temperatures lower than 20 °C (Figures 4.6 and 4.7), we conclude that this microorganism is not able to compete with PAO at cold and moderate temperatures (below 20 °C). From 20 to 30 °C, *Competibacter* can effectively compete with PAO because of their higher anaerobic acetate uptake rate. However, it has a lower biomass growth rate, requiring a minimum aerobic SRT longer than PAO.

As the enriched *Competibacter* culture was exposed for long-term periods (of at least 3 times the applied SRT) to any of the temperatures of study, the microbial conversions involved in the metabolism of GAO were affected depending upon the temperature changes (from 20 °C to 10, 15, 30, 35 and 40 °C). This led to different steady-state conditions when *Competibacter* was able to adapt and acclimate to the switch in temperature (at 15, 30 and 35 °C). The differences in the short- and long-term temperature dependencies of the maximum acetate uptake rate as well as the need to adjust the required SRT clearly reflect the long-term temperature effects on the microbial conversions of *Competibacter*. While the diverse steady-state conditions shown by *Competibacter* at long-term are a resultant of the equilibrium reached among the different anaerobic and aerobic metabolic conversions (e. g. at 15, 30 and 35 °C), the wash-out of *Competibacter* indicates that certain metabolic processes and conversions were severely affected (or even inhibited) and no equilibrium could be achieved (i. e. at 10 and 40 °C). Even at 10 °C a complete switch in population to an enriched *Accumulibacter* culture was observed. These observations exemplify the importance of the execution of long-term experiments compared to short-term studies where potential biomass adaptations and population changes can not be observed. Nevertheless, short-term studies are required when aiming at studying the physiology of microorganisms without involving biomass adaptations or population changes or when evaluating at short-term the potential exposure of microorganisms to conditions or scenarios different than the usually expected (e.g. sudden changes in the wastewater characteristics or on the environmental or operating conditions).

4.5. Conclusions

In the present study, the long-term temperature effects on an enriched GAO culture (*Competibacter*) were evaluated within a relatively wide temperature range (from 10 to 40 °C) that comprises the operating temperature of most of the municipal and industrial WWTP. The anaerobic stoichiometry of GAO (*Competibacter*) was insensitive to temperature changes from 15 to 35 °C. The anaerobic glycogen conversion appeared to be the limiting metabolic process of GAO at low temperature (10 °C). Operation of EBPR lab-scale reactors at low temperature (e.g. 10 °C) could be used as a strategy to suppress the growth of GAO aiming at obtaining highly enriched PAO cultures. At temperatures lower than 35 °C, the aerobic stoichiometry was also independent of temperature changes. Due to strong increases in the aerobic maintenance requirements and less efficient energy production (oxidative phosphorylation activity), the net aerobic growth of GAO was limited at high temperatures (30 and 40 °C). Short- and long-term estimated anaerobic substrate uptake rates were different. In adapted cultures, the rate is independent upon temperature from 20 to 35 °C. Like previously observed for PAO, the growth rate of GAO is a resultant of the other aerobic metabolic processes and not an intrinsic property of the bacteria. For the whole studied temperature range, the computed minimum aerobic solids retention time of *Competibacter* ($SRT^{AER}_{MIN,GAO}$) was longer than that of PAO ($SRT^{AER}_{MIN,PAO}$), implying that the proliferation of *Competibacter* in EBPR systems could be suppressed through adjusting (e.g. shortening) the applied solids retention time (SRT). Concerning the PAO-GAO competition, *Competibacter* cannot compete with *Accumulibacter* at temperatures below 20 °C because of lower acetate uptake and biomass production rates. Above 20 °C, *Competibacter* can effectively compete with *Accumulibacter* due to a higher acetate uptake rate.

Acknowledgements

The authors acknowledge the National Council for Science and Technology (CONACYT, Mexico) and the Autonomous University of the State of Mexico for the scholarship awarded to Carlos Manuel Lopez Vazquez. Special thanks to the lab-staff from UNESCO-IHE Institute for Water Education.

References

Amann RI (1995) *In situ* identification of microorganisms by whole cell hybridization with rRNA-targeted nucleic acid probes. In: Akkermans ADL, van Elsas JD, de Bruijn FJ, editors. Molecular Microbial Ecology Manual. London: Kluwer Academic Publisher. p 1-15.

APHA/AWWA (1995) Standard methods for the examination of water and wastewater, 19th ed. Port City Press, Baltimore.

Bazin MJ, Prosser JI (2000) Physiological models in microbiology. CRC Series in mathematical models in microbiology. Florida: CRC Press.

Brdjanovic D, van Loosdrecht MCM, Hooijmans CM, Alaerts GJ, Heijnen JJ (1997) Temperature effects on physiology of biological phosphorus removal. ASCE J Environ Eng 123(2):144-154.

Brdjanovic D, Logemann S, van Loosdrecht MCM, Hooijmans CM, Alaerts GJ, Heijnen JJ (1998a) Influence of temperature on biological phosphorus removal: process and molecular ecological studies. Water Res 32(4):1035-1048.

Brdjanovic D, van Loosdrecht MCM, Hooijmans CM, Alaerts GJ, Heijnen JJ. (1998b) Minimal aerobic sludge retention time in biological phosphorus removal systems. Biotech Bioeng 60(3):326-332.

Cech JS, Hartman P (1993) Competition between phosphate and polysaccharide accumulating bacteria in enhanced biological phosphorus removal systems. Water Res 27:1219-1225.

Crocetti GR, Hugenholtz P, Bond PL, Schuler A, Keller J, Jenkins D, Blackall LL. (2000) Identification of polyphosphate-accumulating organisms and design of 16S rRNA-directed probes for their detection and quantitation. Appl Environ Microbiol 66(3):1175–1182.

Crocetti GR, Banfield JF, Keller J, Bond PL, Blackall LL (2002) Glycogen accumulating organisms in laboratory-scale and full-scale wastewater treatment processes. Microbiol 148:3353-3364.

Daims H, Bruhl A, Amman R, Schleifer KH, Wagner M (1999) The domain-specific probe EUB 338 is insufficient for the detection of all bacteria: development and evaluation of a more comprehensive probe set. Syst Appl Microbiol 22:434-444.

Erdal UG, Erdal ZK, Randall CW (2003) The competition between PAO (phosphorus accumulating organisms) and GAO (glycogen accumulating organisms) in EBPR (enhanced biological phosphorus removal) systems at different temperatures and the effects on system performance. Water Sci Technol 47(11):1-8.

Filipe CDM, Daigger GT, Grady Jr CPL (2001) A metabolic model for acetate uptake under anaerobic conditions by glycogen-accumulating organisms: stoichiometry, kinetics and effect of pH. Biotechnol Bioeng 76(1):17-31.

Henze M, Gujer W, Mino T, van Loosdrecht MCM (2000) Activated sludge models ASM1, ASM2, ASM2d and ASM3. IWA Scientific and Technical Report No. 9. IWA Task Group on Mathematical Modelling for Design and Operation of Biological Wastewater Treatment. London: IWA Publishing.

Kong YH, Ong SL, Ng WJ, Liu WT (2002) Diversity and distribution of a deeply branched novel proteobacterial group found in anaerobic–aerobic activated sludge processes. Environ Microbiol 4(11): 753–757.

Krishna C, van Loosdrecht MCM (1999a) Effect of temperature on storage polymers and settleability of activated sludge. Water Res 33(10):2374-2382.

Krishna C, van Loosdrecht MCM (1999b) Substrate flux into storage and growth in relation to activated sludge modeling. Water Res 33(14):3149-3161.

Liu WT, Nakamura K, Matsuo T, Mino T (1997) Internal energy-based competition between polyphosphate- and glycogen-accumulating bacteria in biological phosphorus removal reactors-effect of P/C feeding ratio. Water Res 31(6):1430-1438.

Lopez-Vazquez CM, Song YI, Hooijmans CM, Brdjanovic D, Moussa MS, Gijzen HJ, van Loosdrecht MCM (2007) Short-term temperature effects on the anaerobic metabolism of glycogen accumulating organisms. Biotech Bioeng 97(3):483-495.

Lopez-Vazquez CM, Song YI, Hooijmans CM, Brdjanovic D, Moussa MS, Gijzen HJ, van Loosdrecht MCM (2008) Temperature effects on the aerobic metabolism of glycogen accumulating organisms. Biotech Bioeng 101(2):295-306.

Meyer RL, Saunders AM, Blackall LL (2006) Putative glycogen accumulating organisms belonging to *Alphaproteobacteria* identified through rRNA-based stable isotope probing. Microbiology-SGM 152:419–429.

Mino T, van Loosdrecht MCM, Heijnen JJ (1998) Microbiology and biochemistry of the enhanced biological phosphorus removal process. Water Res 32(11):3193-3207.

Murnleitner E, Kuba T., van Loosdrecht MCM, Heijnen JJ (1997) An integrated metabolic model for the aerobic and denitrifying biological phosphorus removal. Biotechnol Bioeng 54(5):434-450.

Oehmen A, Yuan Z, Blackall LL, Keller J (2005) Comparison of acetate and propionate uptake by polyphosphate accumulating organisms and glycogen accumulating organisms. Biotechnol Bioeng 91(2): 162-168.

Oehmen A, Lemos PC, Carvalho G, Yuan Z, Keller J, Blackall LL, Reis MAM. (2007) Advances in enhanced biological phosphorus removal: From micro to macro scale. Wat Res 41(11):2271-2300.

Panswad T, Doungchai A, Anotai J (2003) Temperature effect on microbial community of enhanced biological phosphorus removal system. Water Res 37:409-415.

Reichert P (1994) AQUASIM - a tool for simulation and data analysis of aquatic systems. Water Sci Technol 30(2): 21-30.

Satoh H, Mino T, Matsuo T (1994) Deterioration of enhanced biological phosphorus removal by the domination of microorganisms without polyphosphate accumulation. Water Sci Technol 30(6):203-211.

Smolders GJF, van der Meij J, van Loosdrecht MCM, Heijnen JJ (1994) Stoichiometric model of the aerobic metabolism of the biological phosphorus removal process. Biotechnol Bioeng 44:837-848.

Smolders GJF, van der Meij J, van Loosdrecht MCM, Heijnen JJ (1995) A structured metabolic model for anaerobic and aerobic stoichiometry and kinetics of the biological phosphorus removal process. Biotechnol Bioeng 47(3):277-287.

van Loosdrecht MCM, Hooijmans CM, Brdjanovic D, Heijnen JJ (1997) Biological phosphorus removal processes: a mini review. Applied microbiology and biotechnology 48: 289-296.

Whang LM, Filipe CDM, Park JK (2007) Model-based evaluation of competition between polyphosphate- and glycogen-accumulating organisms. Water Res 41(6):1312-1324.

Whang LM, Park JK (2006) Competition between polyphosphate- and glycogen-accumulating organisms in enhanced biological phosphorus removal systems: effect of temperature and sludge age. Water Environ Res 78(1):4-11.

Wong MT, Tan FM, Ng WJ, Liu WT (2004) Identification and occurrence of tetrad-forming Alphaproteobacteria in anaerobic–aerobic activated sludge processes. Microbiology-SGM 150:3741–3748.

Zeng RJ, van Loosdrecht MCM, Yuan Z, Keller J (2002) Proposed modifications to metabolic model for glycogen accumulating organisms under anaerobic conditions. Biotechnol Bioeng 80(3):277-279.

Zeng RJ, van Loosdrecht MCM, Yuan Z, Keller J (2003) Metabolic model for glycogen-accumulating organisms in anaerobic/aerobic activated sludge systems. Biotechnol Bioeng 81(1):92-105.

Appendix A. Cycles measured in the different studied temperatures.

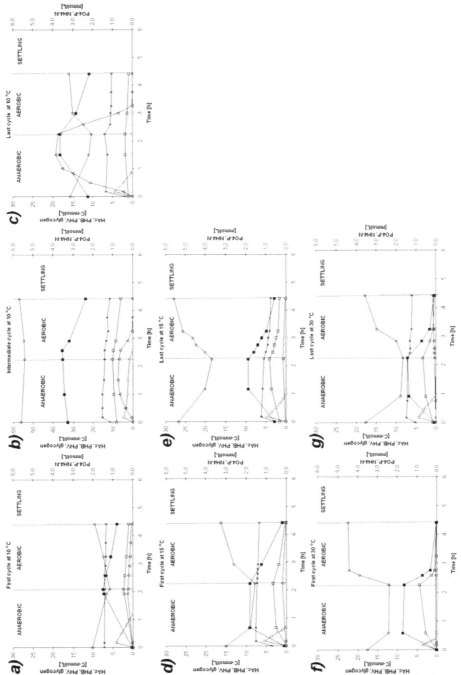

Figure A. Carbon and orthophosphate concentrations measured during the SBR first, intermediate and long-term cycles at 10, 15 and 30 °C: acetate (◊), glycogen (○), PHB (■), PHV (□) orthophosphate (Δ) and ammonium (ж).

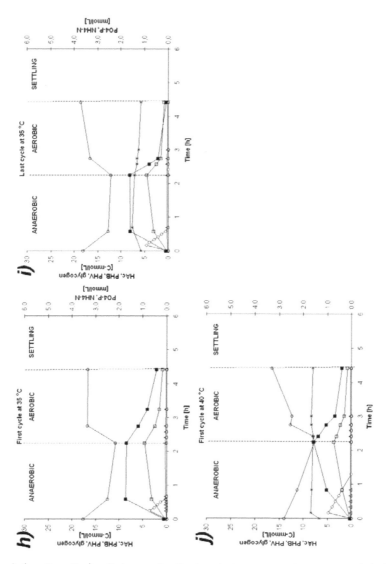

Figure A (continuation). Carbon and orthophosphate concentrations measured during the SBR first and long-term cycles at 35 and 40 °C: acetate (◊), glycogen (○), PHB (■), PHV (□), orthophosphate (Δ) and ammonium (ж).

Appendix B. Determination of the minimal SRT of GAO

Following a similar approach like Brdjanovic et al. (1998b), the minimal aerobic SRT of GAO (*Competibacter*) was calculated using the following expression:

$$SRT_{MIN,GAO}^{AER} = \frac{\left[Y_{PHA,HAc} / Y_{X,HAc}\right] \cdot \left(t_{AER} / 24\right)}{f_{PHA}^{MAX} - \left[\left(f_{PHA}^{MAX}\right)^{1/3} - \dfrac{1}{3} k_{PHA}^{MAX} \cdot t_{AER}\right]} \tag{4.3}$$

where,

$SRT_{MIN,GAO}^{AER}$ is the minimum required aerobic SRT, in days.

$Y_{PHA.HAc}$ is the PHA produced per HAc consumed in the anaerobic phase, 1.86 C-mol/C-mol (Zeng et al., 2003).

$Y_{X.HAc}$ is the net biomass production per HAc consumed per cycle observed in the long-term experiments from this study, in C-mol/C-mol units.

t_{AER} is the length time of the 2.25 h aerobic phase, in days.

f_{PHA}^{MAX} is the maximum PHA fraction present in the biomass, assumed to be around 0.55 C-mol/C-mol (Brdjanovic et al., 1998b; Lopez-Vazquez et al., 2007, 2008).

k_{PHA}^{MAX} $0.606 \cdot k_{PHA,20}^{MAX} \cdot \theta_1^{(T-20)} \cdot \left[1 - \theta_2^{(T-40)}\right]$, this study, see Figure 4.5c.

The $SRT_{MIN,GAO}^{AER}$ were calculated for the whole studied temperature range (from 10 to 40 °C) and are shown in Figure 4.7. The $SRT_{MIN,PAO}^{AER}$ computed by Brdjanovic et al. (1998b) are also included in Figure 4.7.

5

Chapter

Factors affecting the occurrence of PAO and GAO at full-scale EBPR systems

Content

This chapter has been published as:

    Lopez-Vazquez CM, Brdjanovic D, Hooijmans CM, Gijzen HJ, van Loosdrecht MCM (2008) Factors affecting the occurrence of phosphorus accumulating organisms (PAO) and glycogen accumulating organisms (GAO) at full-scale enhanced biological phosphorus removal (EBPR) wastewater treatment plants. *Water Res* 42(10-11):2349-2360.

Abstract

The influence of operating and environmental conditions on the microbial populations of the Enhanced Biological Phosphorus Removal (EBPR) process at seven full-scale municipal activated sludge wastewater treatment plants (WWTPs) in The Netherlands was studied. Data from the selected WWTPs concerning process configuration, operating and environmental conditions were compiled. The EBPR activity from each plant was determined by execution of anaerobic-anoxic-aerobic batch tests using fresh activated sludge. Fractions of *Accumulibacter* as potential PAO and *Competibacter*, *Defluviicoccus*-related microorganisms and *Sphingomonas* as potential GAO were quantified using Fluorescence *in situ* Hybridization (FISH). The relationships among plant process configurations, operating parameters, environmental conditions, EBPR activity and microbial populations fractions were evaluated using a statistical approach. A well-defined and operated denitrification stage and a higher mixed liquor pH value in the anaerobic stage were positively correlated with the occurrence of *Accumulibacter*. A well-defined denitrification stage also stimulated the development of Denitrifying Phosphorus Accumulating Organisms (DPAO). A positive correlation was observed between *Competibacter* fractions and organic matter concentrations in the influent. Nevertheless, *Competibacter* did not cause a major effect on the EBPR performance. The observed *Competibacter* fractions were not in the range that would have led to EBPR deterioration. Likely, the low average sewerage temperature (12 ± 2 °C) limited their proliferation. *Defluviicoccus*-related microorganisms were only seen in negligible fractions in a few plants (< 0.1 % as EUB) whereas *Sphingomonas* were not observed.

5.1. Introduction

In order to identify key factors in the PAO-GAO competition, different lab-scale studies have been carried out providing insight knowledge that might ultimately lead to an improved stability and optimization of the EBPR process. Factors such as type of influent carbon source (e.g. acetate, propionate, etc.) (Sudiana *et al.*, 1999; Oehmen *et al.*, 2004, 2005), influent phosphorus to carbon ratio (P/C) (Liu *et al.*, 1997; Schuler and Jenkins, 2003), pH (Filipe *et al.*, 2001b; Schuler and Jenkins, 2002) and temperature (Whang and Park, 2006; Lopez-Vazquez *et al.*, 2007) have been pointed out as determining factors to understand the PAO-GAO competition.

The importance and overall benefits of Denitrifying Phosphorus Accumulating Organisms (DPAO) in activated sludge systems has also been widely recognized (Kuba *et al.*, 1996b). Nevertheless, the factors that may influence the occurrence of DPAO are not well understood and discrepancies were observed when the DPAO's anoxic phosphorus removal activity, generally defined as a fraction of PAO's aerobic activity, was calculated. Kerrn-Jespersen and Henze (1993), Kuba *et al.* (1996a, 1997a, 1997b), Meinhold *et al.* (1999) and Hu *et al.* (2002) reported different DPAO's anoxic activities that varied from 15 to even 100 % of the aerobic P-removal activity. Moreover, some other questions arise related to the potential contribution of GAO to denitrification processes at full-scale systems. Zeng *et al.* (2003b) described that a lab-enriched GAO culture was also able to denitrify. However, the final product of denitrification was nitrous oxide (N_2O), considered an undesirable greenhouse gas, rather than dinitrogen gas (N_2). This fact might create environmental concerns in cases when Denitrifying Glycogen-Accumulating Organisms (DGAO) play a major role in denitrification at full-scale plants.

It should be underlined that practically all studies have been carried out at lab-scale whereas comparative data and observations from full-scale studies are still limited. Therefore, there is a clear need to verify and extend the findings of laboratory-based studies to full-scale operating plants (Seviour *et al.*, 2003). Since there is a gap between conclusions drawn from lab- and full-scale research the objective of the present paper is to evaluate which and how environmental and operating conditions as well as wastewater characteristics may influence the occurrence of PAO, DPAO, GAO and DGAO at full-scale EBPR wastewater treatment plants (WWTPs). Additionally, a particular objective of this study was to evaluate the microbial communities distributions at low sewage temperature (below 20 °C) considering that temperature lower than 20 °C appears to be more favorable for PAO than for GAO (Whang and Park, 2006; Lopez-Vazquez *et al.*, 2007).

In order to achieve the objectives of this study, seven EBPR WWTPs were surveyed in The Netherlands during winter (November 2005 to April 2006). For the period of study, data regarding the process plant configuration and operating and environmental conditions were collected. The EBPR activity was evaluated by executing anaerobic-anoxic-aerobic batch tests using fresh activated sludge from each WWTP. Fluorescence *in situ* Hybridization (FISH) was used to determine the PAO and GAO fractions present in the treatment plants subjected to this survey. A statistical analysis was performed in order to find the most prominent relationships among operational and environmental data, EBPR activity and quantified PAO and GAO populations.

5.2. Materials and methods

The survey was carried out from November 2005 to April 2006 under (Dutch) winter conditions with an average sewage temperature of around 12 °C. Visits to the WWTP were undertaken in the last months of the survey (March and April 2006) assuming that the microbial populations were already stable after the seasonal temperature change and before the atmospheric temperature increased due to the spring-summer period.

Seven Dutch municipal wastewater treatment plants were selected with the aim of covering the range of the most common EBPR treatment systems used in The Netherlands. The study comprised four treatment plants operated with a Phoredox process plant configuration (Katwoude, Hoek van Holland, Venlo and Waarde WWTPs), two with a modified UCT process (Hardenberg and Deventer WWTPs) (BCFS^{TM}, van Loosdrecht *et al.*, 1998) and one plant (Haarlem Waarderpolder WWTP) using an EBPR sidestream P-stripping process.

5.2.1. Data collection

For the period of study, information related to plant process configurations, operational, control as well as design aspects of the treatment plants were directly obtained by reviewing documentation, interviewing practitioners and plant operators and visiting WWTPs facilities.

Average influent and effluent characteristics were determined based on data provided by plant operators by calculating the arithmetic mean and standard deviation concerning the period of study. However, in order to complete the required data and because certain parameters (such as Volatile Fatty Acids, VFA), were not measured on a regular basis at the plants, a sampling campaign was designed and implemented during visits. Additional grab samples were collected at the influent, start and end of each biochemical stage (anaerobic, anoxic and aerobic) and at the effluent of the secondary clarifier. Samples for the determination of soluble parameters, such as orthophosphate, were immediately filtered through 0.45 μm pore size disposable filters. After collection, all samples were kept on ice or in a fridge at 4 °C till the execution of the corresponding analyses (< 24 h).

5.2.2. Batch tests

Activated sludge was collected during the visits to the different surveyed WWTP. Activated sludge was not collected during peak events, rainy days or after raining events aiming at minimizing the effects of influent dynamic conditions on microbial activities. Collected sludge was kept on ice during transportation and, after arrival to the laboratory and before the execution of the batch activity tests, at 4 °C in a temperature-controlled room. During transportation and storage, activated sludge was not aerated since anaerobic starvation conditions have a lower negative effect on EBPR communities than aerobic conditions (Lopez *et al.*, 2006). Batch tests to evaluate the EBPR activities from investigated WWTPs were performed using a 1.5 L and a 0.5 L double-jacketed laboratory fermentors. 1 L of activated sludge collected at the end of the aerobic tank from each WWTP was used to carry out the corresponding anaerobic-anoxic-aerobic batch tests within 24 h after collection. Both fermentors were operated at controlled temperature of 20 ± 0.5 °C and pH 7.0 ± 0.1. N_2 gas was supplied at a flowrate of 30 L/h during the anaerobic and anoxic phases to avoid oxygen intrusion, while 60 L/h of air were provided during the aerobic phase. Prior to execution of the activity tests, sludge was aerated for at least 1 h in order to acclimatize it to the lab batch reactor conditions. Batch tests were executed to determine: (a) maximum acetate uptake rate, (b) anaerobic P-release rate, (c) anaerobic P/HAc ratio, (d) aerobic P-uptake rate, (e) anoxic P-uptake rate and (f) anoxic/aerobic P-uptake rate ratio. The description of the experimental procedure to determine the anaerobic-anoxic-aerobic EBPR activities can be found elsewhere (Kuba *et al.*, 1997a, 1997b; Wachtmeister *et al.*, 1997).

During the execution of the batch tests, acetate (HAc), orthophosphate (PO_4^{3-}-P), nitrate (NO_3^--N), mixed liquor suspended solids and volatile suspended solids (MLVSS and MLVSS, respectively) were measured. Synthetic medium used to carry out the batch tests was prepared as described by Smolders *et al.* (1994) but diluted in demineralized water in order to reach an initial acetate concentration at the beginning of the batch tests of 40 mg/L.

5.2.3. Analyses

PO_4^{3-}-P was determined by the ascorbic acid method. All analyses, including MLSS and MLVSS determination, were performed in accordance with Standard Methods (A.P.H.A., 1995). Meanwhile, NO_3^--N determination was carried out with a Dionex ICS-1000 Ion Chromatograph (IC) (Dionex, The Netherlands) equipped with a AS14A analytical column, AG14A guard column and using a 8 mmol Na_2CO_3 plus 1 mmol $NaHCO_3$ solution (3.39 g $NaCO_3$ plus 0.336 g $NaHCO_3$ in 4 mL of Milli-Q water) as eluant at a flowrate of 0.7 mL/min. Temperature of the oven and cell (detector) was 40 °C. Detection limit was 0.01 mg/L for a 10 μL sample injection. Volatile Fatty Acids (VFA) fractions as acetate (HAc), propionate (HPr), butyrate (HBr) and valerate (HVr) were determined through Gas Chromatography (GC) using a Chrompack CP 9001 gas chromatograph (Varian-Chrompack, Bergen on Zoom, The Netherlands) equipped with a flame ionization detector (FID) operated at 300 °C, splitter at 175 °C and a wall coated open tubular (WCOT) fused silica capillary column from Chrompack for free fatty acids phase chemically bound (CP-FFAP CB) of 25 m × 0.53 mm × 1 mm. Helium was used as carrier gas at 35 kPa flow pressure. **The oven temperature was kept constant at 115 °C for 13 min. The detection limit was around 5 mg/L for a 1 μL sample injection.**

5.2.4. FISH analysis

For the quantification of the microbial communities, samples for Fluorescence *in situ* Hibridization (FISH) from the seven different treatment plants were collected. FISH analyses were performed as described in Amann (1995). FISH probes used in this study were EUBMIX (mixture of probes EUB 338, EUB338-II and EUB338-III) for *Bacteria* (Amman *et al.*, 1995; Daims *et al.*, 1999); PAOMIX (equal amounts of PAO462, PAO651 and PAO 846) to target the *Betaproteobacteria Accumulibacter spp.* (Crocetti *et al.*, 2000); GAOMIX (equal amounts of GAOQ989 plus GB_G2) for the *Betaproteobacteria Competibacter spp.* (Crocetti *et al.*, 2002; Kong *et al.*, 2002); DF1MIX (TFO_DF218 plus TFO_DF218) for the *Alphaproteobacteria* from Cluster 1 *Defluviicoccus spp.* (Wong *et al.*, 2004); DF2MIX (DF988, DF1020 with helper probes H966 and H1038 in equal amounts) for Cluster 2 *Defluviicoccus spp.* (Meyer *et al.*, 2006), and SBR9-1a for the *Alphaproteobacteria Sphingomonas spp.* (Beer *et al.*, 2004). Large aggregates were avoided by mild sonication (5W, 30 s). FISH preparations were visualized with a Zeiss Axioplan 2 microscope.

Quantifications of population distributions were carried out using the MATLAB image processing toolbox (The Mathworks, Natick, MA) as described elsewhere (Lopez-Vazquez *et al.*, 2007). Around twenty separate randomly chosen images were evaluated with final results reflecting the average fractions of *Accumulibacter*, *Competibacter*, Cluster 1 *Defluviicoccus spp.*, Cluster 2 *Defluviicoccus spp.* and *Sphingomonas* present in corresponding activated sludge samples. Microbial populations fractions were expressed as percentage of EUB. The standard error of the mean (SEM) was calculated as the standard deviation of the area percentages divided by the square root of the number of images analysed.

5.2.5. Statistical analysis

In the current survey, correlations among thirty different process, design, operational and environmental parameters were evaluated. The studied parameters comprised results from the population quantification analyses, operational data from the WWTPs and results of the EBPR lab-activity tests.

The studied parameters were grouped in six major categories: (1) *Accumulibacter* and *Competibacter* fractions: expressed as *Accumulibacter*/EUB, *Competibacter*/EUB and *Competibacter*/*Accumulibacter* ratios; (2) Design aspects: nominal anaerobic, anoxic and aerobic hydraulic retention times (HRT) and anoxic/aerobic HRT ratios; (3) Influent characteristics: BOD_{INF}, COD_{INF}, BOD_{INF}/COD_{INF} ratio, VFA_{INF}, VFA_{INF}/COD_{INF} ratio, HPr_{INF}/VFA_{INF}, total P_{INF}, total N_{INF}, total P_{INF}/VFA_{INF} and total P_{INF}/BOD_{INF} ratio; (4) pH and water temperature measured in the anaerobic tank; (5) Effluent characteristics: total P_{EFF}, PO_4^{3-}-P_{EFF}, NH_4^--N_{EFF}, NO_X^--N_{EFF} and total N_{EFF}; and, (6) EBPR activity observed in the anaerobic-anoxic-aerobic activity tests: maximum anaerobic HAc uptake rate, anaerobic P-release rate, anaerobic P/HAc ratio, anoxic P-uptake rate, aerobic P-uptake rate, and anoxic/aerobic P-uptake rate ratio.

Collected data, concerning 7 treatment plants and 30 parameters, were compiled and organized in an 7 x 30 matrix. In order to find the strength of relationship for each pair of parameters, a correlation coefficient matrix (R) was generated. The correlation coefficient matrix was computed as follows:

$$R_{i,j} = \frac{C(i,j)}{\sqrt{C(i,i)C(j,j)}} \qquad (5.1)$$

Where $R_{i,j}$ indicates the correlation coefficient for the pair of parameters i and j being i and j any of the possible 30 different parameters; $C(i,j)$ the covariance associated to the parameters; and, $C(i,i)$ and $C(j,j)$ represent the variances of studied parameters. This led to a symmetric 30 x 30 matrix.

As a second step, a 30 x 30 matrix of p-values (P), for testing the hypothesis of no correlation was calculated. Understanding each p-value as the probability of getting a correlation as large as the observed value by random chance, when the true correlation is zero. If $P(i,j)$ is smaller than 0.05, then the correlation $R(i,j)$ is significant. Calculations of the matrices of correlation coefficients and p-values were carried out using MATLAB (The Mathworks, Natick, MA).

5.3. Results

5.3.1. Operational data from wastewater treatment plants

Selected characteristics of process configurations of the seven treatment plants of this survey are shown in Table 5.1. Despite the different applied configurations, practically all WWTPs were designed with emphasis on the creation of plug-flow conditions via separate units or compartmentalization of tanks. At 6 out of 7 plants, there was no supplementary chemical dosing for phosphorus precipitation. Primary settling is not a common practice in EBPR plants in The Netherlands.

Table 5.1. Characteristics of the wastewater treatment plants evaluated in this survey.

WWTP	Phosphorus Removal Configuration	Main configuration of the tanks	Pre-settling	Anaerobic selector	Denitrification phase respect to anaerobic phase	Supplementary Chemical Dosing for P-precipitation	Nominal Hydraulic Retention Times [a]		
							Anaerobic [h]	Anoxic [h]	Aerobic [h]
Hardenberg	Modified UCT process	Carrousel-type	No	Continuous	Separate pre-denitrification	No	6.0	12.0	31.9
Deventer	Modified UCT process	Carrousel-type	Yes	Continuous	Separate pre-denitrification	No	2.7	7.8	16.4
Katwoude	Phoredox	Carrousel-type	No	Separate	Separate pre-denitrification	Yes [b]	0.5	5.2	46.4
Hoek van Holland	Phoredox	Carrousel-type	Yes	Continuous	Separate pre-denitrification	No	3.1	5.5	15.9
Venlo	Phoredox	Carrousel-type	No	Continuous	No separate pre-denitrification [c]	No	2.8	0.0 [c]	7.4
Waarde	Phoredox	Completely mixed reactor	No	Continuous	No separate pre-denitrification [c]	No	1.7	0.0 [c]	13.9
Haarlem Waarderpolder	Sidestream [d]	Carrousel-type	Yes	Separate Side-stream	Separate pre-denitrification	No	11.5	10.0	10.0

WWTP characteristics and parameters

(a) Calculated considering average influent flow-rates for the period of the study and volume of the tanks.

(b) Supplementary chemical dosing takes place when effluent PO_4^{3-}-P concentration exceeds a maximum value of 0.75 mg/L.

(c). Not defined denitrification tanks, intermittent anoxic conditions are created by inactivating aeration systems when effluent NH_4-N concentration is lower than 0.50 mg/L but avoiding non-aerated periods longer than 30 minutes.

(d) Side-stream P-stripping process for biological phosphorus removal (Brdjanovic *et al.*, 2000).

At 5 out of 7 WWTPs investigated, the intrusion of oxygen, nitrate or nitrite in anaerobic tanks tended to be minimized by the implementation of separate pre-denitrification tanks. Only in two particular cases (Venlo and Waarde WWTPs) well-defined separate anoxic phases were not present. In these two plants denitrification takes place in the aerobic reactors, through creation of intermittent anoxic phases, by inactivating the aeration system when effluent NH_4^+-N concentration becomes lower than 0.5 mg/L but avoiding idle phases longer than 0.5 h. Thus, the anoxic phase length depended on the aerobic phase length, which was determined by the wastewater influent load. This practice made it difficult to determine the anoxic phase length. Moreover, under the cold climate conditions that this study was carried out (winter season), a longer aerobic stage is required in order to achieve a satisfactory nitrification activity. Consequently, for the study period, it was considered that these WWTPs were hardly able to create an anoxic stage and, for the purposes of this study, it was assumed that their nominal anoxic HRT was zero.

A summary of compiled average operational data for the period of study is shown in Table 5.2. Based on interviews with operators, 5 out of 7 WWTPs received industrial wastewater discharges during the period of evaluation. However, they were mostly intermittent and comprised less than 25 % of total influent flows. Therefore, it can be assumed that wastewater streams treated by these plants were mostly domestic.

Both wastewater influent quality and quantity varied considerably from one treatment plan to another, which made it difficult to find similarities between obtained operational data. Herewith, certain selected characteristics are presented: (a) mean water temperature and pH within the study period were around 12 ± 2 °C and 6.9 ± 0.3, respectively, (b) VFA_{INF} comprised, in average, about 18 % of total COD_{INF}, (c) average propionate (HPr_{INF}) to VFA_{INF} ratio was 0.13 mg/mg although a high variability was observed from plant to plant as indicated by the large standard deviation ($\pm$ 0.17 mg/mg), and (d) mean total P_{INF}/VFA_{INF} was 0.12 ± 0.05 mg P/mg VFA.

It was noteworthy that all wastewater treatment plants included in this survey are achieving low average effluent concentrations concerning nutrients (6 and 0.5 mg/L for total nitrogen and total phosphorus, respectively) that met the corresponding Dutch effluent standards for nitrogen and phosphorus with a considerable margin (total N < 10 mg/L and total P < 1 or 2 mg/L, depending on plant capacity and location).

Table 5.2. Average operational data in the period of study of the wastewater treatment plants evaluated in this survey.

Characteristics and parameters	Wastewater treatment plants						
	Hardenberg	Deventer	Katwoude	Hoek van Holland	Venlo	Waarde	Haarlem Waarderpolder
Period	Nov 2005 to Dec 2005	Nov 2005 to Jan 2006	Nov 2005 to Mar 2006	Nov 2005 to Feb 2006	Nov 2005 to Dec 2005	Nov 2005 to Dec 2005	Mar 2006
Temperature [°C]	9	14	11	13	15	12	13
pH anaerobic tank	7.2	7	7.2	7.2	6.9	6.6	6.4
Influent parameters							
Average Influent Flow [m³/h]	253±74	366±128	342±12	917±183	2975	448±211	1536±448
BOD [mg/L]	308±41	211±42	251±81	110±39	191±92	136±57	64±55
COD [mg/L]	774±68	367±92	580±156	290±87	510±221	420±121	327±143
VFA [mg/L]	110	41	102	45	91	175	33
HAc [mg/L]	88	41	91	39	77	93	28
HPr [mg/L]	22	n.d.	11	5	n.d.	82	n.d.
HPr/VFA [mg/L]	0.20	0.00	0.11	0.11	0.00	0.47	0.00
VFA/COD [mg/L]	0.14	0.11	0.18	0.15	0.18	0.42	0.10
N total [mg/L]	80±17	61±12	59±12	72±14	47±17	37±14	36±17
P total [mg/L]	11±1	7±1	9±2	7±1	10±4	6±3	6±2.3
PO_4^{3-}-P [mg/L]	-	5±1	-	6	6±3	4±3	-
P/VFA ratio [mg/mg]	0.10	0.17	0.09	0.14	0.11	0.03	0.17
P/BOD ratio [mg/mg]	0.04	0.03	0.04	0.06	0.05	0.04	0.09
Effluent parameters							
BOD [mg/L]	2	-	3	-	1	2	3
COD [mg/L]	40±7	36±6	48±18	67±13	41±11	49±20	26±5
NH4-N [mg/L]	0.1±0.1	1.5±0.1	1.7±2.9	1±1	0.6±0.4	0.5±0.2	3.0±1.5
NOx-N [mg/L]	3.0±1.3	5.0±1.5	1.9±1.4	3.6±3.6	2.9±1.1	5.1±2.7	3.0±1.1
N total [mg/L]	5.2±1.5	6.6±1.3	5.9±4.9	5.4±1.1	4.7±1.5	7.1±2.8	4.8±0.9
P total [mg/L]	0.6±0.2	0.9±0.1	0.4±0.2	0.5±0.1	0.3±0.1	0.6±0.5	0.3±0.1
PO_4^{3-}-P [mg/L]	0.4±0.1	0.7±0.1	0.3±0.3	0.2±0.2	0.0	0.4±0.3	0.1±0.1

-. Not measured.
n.d. - Not detected.

5.3.2. EBPR activity observed in batch tests

Results of the anaerobic-anoxic-aerobic activity tests are shown in Table 5.3. Average acetate uptake rates, P-release rates and P/HAc ratios were 16 ± 5 mg HAc/gVSS/h, 13 ± 4 mg PO_4^{3-}-P/gVSS/h and 0.40 ± 0.04 mg PO_4^{3-}-P/mgHAc, respectively. Considerable fluctuations were observed on the aerobic and anoxic P-uptake rates (average rates were 10 ± 4 mg PO_4^{3-}-P/gVSS/h and 3 ± 2 mg PO_4^{3-}-P/gVSS/h, respectively) that led to anoxic/aerobic P-uptake rate ratios that fluctuated from 48 % to lower values such as 9 % and even zero (average anoxic/aerobic P-uptake rate ratio was 24 ± 16 %). As presented in Table 5.3, Kuba *et al.* (1997a, 1997b), Meinhold *et al.* (1999) and Brdjanovic *et al.* (2000) also observed a great variability in the rates and ratios measured in other full-scale systems.

5.3.3. Quantification of microbial populations

Quantification of the different microbial populations (*Accumulibacter, Competibacter, Defluviicoccus-related* microorganisms and *Sphingomonas*) from the seven different WWTP is presented in Table 5.4. *Accumulibacter*, in average, comprised around 9.2 ± 2.1 % of total biomass (as EUB), *Competibacter* about 1.7 ± 0.4 % (as EUB) leading to a *Competibacter*/*Accumulibacter* ratio of 18 ± 4 %. Negligible fractions (< 0.1 %) of *Defluviicoccus*-related microorganisms Cluster 1 and 2 were only observed in three plants (Hardenberg –HRD–, Venlo –VEN– and Haarlem Waarderpolder –HW–). *Sphingomonas* were not detected in any of treatment plants of this survey.

Regarding the *Accumulibacter* and *Competibacter* fractions, two different trends were observed (Figure 5.1). In four treatment plants (Hoek van Holland –HvH–, Hardenberg –HRD–, Katwoude –KAT– and Deventer – DEV–) average *Accumulibacter* fraction was 11.6 ± 2.3 % while in the other plants 6.0 ± 1.9 % (Waarde –WAR–, Venlo –VEN– and Haarlem Waarderpolder –HW–). This may suggest that plant configuration and operating conditions of these two groups of plants had a different effect or influence on the development of *Accumulibacter*. These values were in the range observed in previous full-scale studies for *Accumulibacter* populations (7 – 22 %) (Zilles *et al.*, 2002; Saunders *et al.*, 2003; Kong *et al.*, 2004; Chua *et al.*, 2006; Tykesson *et al.*, 2006). Differences among the evaluated WWTPs for the *Competibacter* populations were not observed. However, lower *Competibacter* fractions were found (1.7 ± 0.4 %) than in previous full-scale reports (1 – 12 %, Saunders *et al.*, 2003; 10 – 31 %, Wong *et al.*, 2005; 5 – 10 %, Tykesson *et al.,* 2006; 0 – 6 %, Kong *et al.*, 2006).

Table 5.3. Summary of the anaerobic-anoxic-aerobic activity tests results carried out with activated sludge from the different wastewater treatment plants evaluated in this study.

WWTP	HAc uptake rate [mg/gVSS/h]	P-release rate [mg/gVSS/h]	Anaerobic P/HAc ratio [mg/mg]	Aerobic P-uptake rate [mg/gVSS/h]	Anoxic P-uptake rate [mg/gVSS/h]	Anoxic/Aerobic P-uptake rate [%]
Hardenberg	21.9	17.4	0.38	19.2	5.9	31
Deventer	19.2	9.6	0.45	9.0	2.1	23
Katwoude	13.5	11.1	0.33	8.0	1.9	23
Hoek van Holland	21.3	20.9	0.38	9.1	4.4	48
Venlo	11.2	10.6	0.38	6.2	0.6	9
Waarde	14.0	13.6	0.45	6.3	0.0	0
Haarlem Waarderpolder	9.0	10.4	0.40	9.8	3.3	34
Other studies						
Haarlem Waarderpolder (Brdjanovic et al., 2000)	23	6	0.31	2.2	1.7	80
Holten (Kuba et al., 1997a,b)	47	16	0.50	13	6	46
Geneműiden (Kuba et al., 1997b)	7 – 31	5 – 9	0.40	4 – 6	1.2 - 1.6	20 - 40
Pilot-scale BIODENIPHO (Meinhold et al., 1999)[a]	n. r.	n. r.	n. r.	4	2	54

n. r.: No reported

(a) Anoxic and aerobic P-uptake rates expressed as mg/gSS/h.

Table 5.4. Microbial populations fractions observed in this study.

WWTP	*Accumulibacter* [%]	*Competibacter* [%]	*Defluviicoccus* cluster 1 [%]	*Defluviicoccus* cluster 2 [%]	*Sphingomonas* [%]
HvH	16.4 ± 2.8	0.6 ± 0.2	n.d	n.d	n.d.
HRD	10.6 ± 1.4	3.2 ± 0.5	n.d.	A few cells [a]	n.d.
KAT	10.0 ± 2.6	2.9 ± 0.6	n.d.	n.d.	n.d.
DEV	9.5 ± 2.3	1.9 ± 0.4	n.d.	n.d.	n.d.
WAR	6.2 ± 1.2	1.0 ± 0.5	n.d.	n.d.	n.d.
VEN	6.1 ± 2.5	2.3 ± 0.3	A few cells [a]	A few cells [a]	n.d.
HW	5.7 ± 2.2	0.4 ± 0.1	A few cells [a]	A few cells [a]	n.d.
Average	9.2 ± 2.1	1.7 ± 0.4	n.d.	n.d.	n.d.

nd. No detected.
[a] Only a few cells were observed (fraction < 0.1 %).

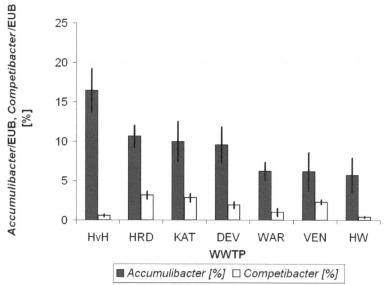

Figure 5.1. Fractions of PAO (*Accumulibacter*) and GAO (*Competibacter*) populations in the wastewater treatment plants of the present study related to total biomass (as EUB). Bars indicate the standard error of the mean (SEM).

5.3.4. Data compilation and significant correlations among variables

Thirty different parameters comprising (a) *Accumulibacter* and *Competibacter* fractions; (b) design and operating aspects of the different

WWTPs; and, (c) results from EBPR activity tests were analyzed. Population fractions of *Defluviicoccus*-related organisms and *Sphingomonas* were not included in the statistical analysis since these microorganisms were only detected in practicably negligible percentages in three plants (Table 5.4). Correlations among 30 different parameters were investigated. The coefficients of correlation (R) and determination (R^2) as well as the significance level of each correlation (p) were also calculated.

Thirty significant correlations were found ($R^2 > 0.50$ and $p < 0.05$), thirteen of them were ranked as highly significant ($R^2 > 0.70$, $p < 0.01$) (Table 5.5). Certain correlations such as *Competibacter* fraction *vs.* *Competibacter/Accumulibacter* fractions ratio or total P_{EFF} *vs.* PO_4^{3-}-P_{EFF} (correlations No. 5 and 27) were expected based on assumed close dependencies between these parameters. Similarly, other expected correlations on the basis of the wastewater characteristics and biomass activity were also confirmed (correlations No. 15 - 19, 21, 23 and 30: BOD_{INF} *vs.* COD_{INF}, BOD_{INF} *vs.* total P_{INF}, COD_{INF} *vs.* total P_{INF}, BOD_{INF}/COD_{INF} *vs.* total P/BOD_{INF}, VFA_{INF} *vs.* VFA_{INF}/COD_{INF} ratio, VFA_{INF} *vs.* total P_{INF}/VFA_{INF}, VFA_{INF}/COD_{INF} *vs.* total P_{INF}/VFA_{INF}, and Anoxic P-uptake rate *vs.* Anoxic/Aerobic P-uptake rates ratio). These results justified the use of such statistical method in this study for the evaluation of the relationships among the selected parameters.

Statistical analysis indicated that the occurrence of *Accumulibacter* was significantly correlated ($R^2 > 0.50$, $p < 0.05$) with (i) the pH level measured in anaerobic tanks, (ii) HAc uptake rate and (iii) P-release rates (Table 5.5, correlations No. 2 - 4). Interestingly, also significant correlations were observed between *Accumulibacter* fractions and total N_{INF} (correlation No. 1) suggesting that DPAO may play an important role in the nutrient removal processes at the full-scale WWTPs evaluated in this survey. The significant contribution of DPAO to nitrogen removal seems to be confirmed considering that operational conditions and parameters primarily related to nitrogen removal showed to have a significant correlation with the activity of *Accumulibacter* (correlations No. 11 - 12, 26, and 28 - 29: Anoxic HRT *vs.* Aerobic P-uptake rate, Anoxic HRT *vs.* Anoxic P-uptake rate, total N_{INF} *vs.* HAc uptake rate, total P_{EFF} *vs.* NOx-N_{EFF}, and Aerobic-P uptake rate *vs.* Anoxic P-uptake rate).

Table 5.5. Significant (p < 0.05) and highly significant correlations (p < 0.01) found among the different studied parameters.

No.	R [i,j]	Parameter i	vs.	Parameter j	p < 0.05	p < 0.01	R	R²
1	R [1,15] = R [15,1]	Accumulibacter fraction	vs.	Total N inf	0.029		0.81	0.65
2	R [1,23] = R [23,1]	Accumulibacter fraction	vs.	pH anaerobic tank	0.050		0.75	0.57
3	R [1,25] = R [25,1]	Accumulibacter fraction	vs.	HAc uptake rate	0.047		0.76	0.58
4	R [1,26] = R [26,1]	Accumulibacter fraction	vs.	P-release rate	0.041		0.78	0.60
5	R [2,3] = R [3,2]	Competibacter fraction	vs.	Competibacter / Accumulibacter fraction	0.001	0.001	0.96	0.92
6	R [2,8] = R [8,2]	Competibacter fraction	vs.	BOD inf	0.004	0.004	0.92	0.84
7	R [2,9] = R [9,2]	Competibacter fraction	vs.	COD inf	0.026		0.81	0.66
8	R [3,8] = R [8,3]	Competibacter / Accumulibacter fraction	vs.	BOD inf	0.016		0.85	0.72
9	R [3,9] = R [9,3]	Competibacter / Accumulibacter fraction	vs.	COD inf	0.027		0.81	0.66
10	R [4,7] = R [7,4]	Anaerobic HRT	vs.	Anoxic/Aerobic HRT ratio	0.007	0.007	0.89	0.80
11	R [5,28] = R [28,5]	Anoxic HRT	vs.	Aerobic P-uptake rate	0.025		0.82	0.67
12	R [5,29] = R [29,5]	Anoxic HRT	vs.	Anoxic P-uptake rate	0.016		0.85	0.72
13	R [7,11] = R [11,7]	Anoxic/Aerobic HRT ratio	vs.	VFA inf	0.045		-0.87	0.59
14	R [7,16] = R [16,7]	Anoxic/Aerobic HRT ratio	vs.	Total P/VFA inf	0.025		0.82	0.67
15	R [8,9] = R [9,8]	BOD inf	vs.	COD inf	0.009	0.009	0.88	0.77
16	R [8,14] = R [14,8]	BOD inf	vs.	Total P inf	0.009	0.009	0.88	0.77
17	R [9,14] = R [14,9]	COD inf	vs.	Total P inf	0.005	0.005	0.90	0.81
18	R [10,17] = R [17,10]	BOD/COD inf	vs.	Total P/BOD inf	0.022		-0.83	0.69
19	R [11,12] = R [12,11]	VFA inf	vs.	VFA/COD inf	0.009	0.009	0.88	0.77
20	R [11,13] = R [13,11]	VFA inf	vs.	HPr/VFA inf	0.011		0.87	0.76
21	R [11,16] = R [16,11]	VFA inf	vs.	Total P/VFA inf	0.000	0.000	-0.98	0.96
22	R [12,13] = R [13,12]	VFA/COD inf	vs.	HPr/VFA inf	0.006	0.006	0.90	0.81
23	R [12,16] = R [16,12]	VFA/COD inf	vs.	Total P/VFA inf	0.008	0.008	-0.89	0.79
24	R [13,16] = R [16,13]	HPr/VFA inf	vs.	Total P/VFA inf	0.012		-0.86	0.75
25	R [15,23] = R [23,15]	Total N inf	vs.	pH anaerobic tank	0.009	0.009	0.88	0.77
26	R [15,25] = R [25,15]	Total N inf	vs.	HAc uptake rate	0.008	0.008	0.89	0.78
27	R [18,19] = R [19,18]	Total P eff	vs.	PO₄-P eff	0.005	0.005	0.90	0.82
28	R [18,21] = R [21,18]	Total P eff	vs.	NOx-N eff	0.029		0.80	0.65
29	R [28,29] = R [29,28]	Aerobic P-uptake rate	vs.	Anoxic P-uptake rate	0.013		0.86	0.74
30	R [29,30] = R [30,29]	Anoxic P-uptake rate	vs.	Anoxic/Aerobic P-uptake rates ratio	0.019		0.84	0.70

Competibacter fractions had significant positive correlations with BOD_{INF} and COD_{INF} concentrations (Table 5.5, correlations No. 6 - 9). Taking into account that a satisfactory biological P-removal performance was observed in all the WWTP of the current study, these relationships may potentially suggest that the influent contained a carbon load higher than the required by PAO for P-removal and that any excess of organic matter was only favorable for the development of *Competibacter*.

The relationship presented in Table 5.5 between nominal Anaerobic HRT and Anoxic/Aerobic HRT (correlation No. 10) seemed to be a reflection of the design criteria for nutrient removal at WWTPs since the anaerobic stages increase proportionally to the anoxic stages.

As influent VFA concentrations and fractions increased (VFA_{INF} and VFA/COD_{INF}) higher fractions of propionate were observed (HPr/VFA_{INF}) suggesting that as fermentation processes increased higher fractions of HPr were produced (Table 5.5, correlations No. 20 and 22). Additionally, relationships among parameters involving VFA and total P were also observed (correlations No. 23 and 24: VFA/COD_{INF} *vs.* total P/VFA_{INF} and HPr/VFA_{INF} *vs.* total P/VFA_{INF}). These correlations could also be explained considering the influence of the fermentation processes: if the VFA_{INF} fraction increases, as consequence of the fermentation processes, then the total P/VFA_{INF} ratio decreases since total P is not affected by fermentation.

An explanation, on the basis of the present study, could not be directly found for the next correlations: Anoxic/Aerobic HRT ratio *vs.* VFA_{INF}, Anoxic/Aerobic HRT ratio *vs.* Total P/VFA_{INF} and total N_{INF} *vs.* pH anaerobic tank (Table 5.5, correlations No. 13-14 and 25). Perhaps, the first two relationships are related to the design criteria for nutrient removal assuming that larger anoxic stages enhance nitrogen and phosphorus removal. While Total N_{INF} *vs.* pH anaerobic tank may be a consequence of the denitrification processes that take place in the anoxic stage: if higher N loads are treated, presumably higher NOx-N loads are removed via denitrification. This could increase the mixed liquor pH value in the anoxic tank. Thus, considering the internal recirculations of activated sludge from anoxic to anaerobic stages, this might result in a pH increase in the anaerobic stage as well resulting beneficial for *Accumulibacter*.

5.4. Discussion

5.4.1. PAO and GAO occurrence at full-scale wastewater treatment plants

Higher PAO fractions (*Accumulibacter*) (11.6 ± 2.3 %) were observed at wastewater treatment plants operated with separate pre-denitrification tanks (Hoek van Holland, Hardenberg, Katwoude and Deventer WWTPs) than in plants operated without defined denitrification stages like Waarde and Venlo WWTPs (6.0 ± 1.9 %, Table 5.4). This underlines once again the need of avoiding the presence of oxygen or NOx-N compounds in anaerobic tanks to preventing the consumption of VFA by ordinary heterotrophs.

In the case of Haarlem Waarderpolder WWTP a lower fraction of *Accumulibacter* was present because only a small fraction of mixed liquor is exposed to anaerobic conditions (< 10 %) being derived and treated in a side-stream P-stripping process designed for biological P-removal (Brdjanovic *et al.*, 2000). According to Smolders *et al.* (1996), in a side-stream bio-P process a lower PAO fraction is required resulting in lower carbon requirements. This may explain why smaller PAO fractions were observed in Haarlem Waarderpolder WWTP.

Katwoude WWTP required the use of supplementary chemical dosing to meet the corresponding effluent standards (< 1mg total P/L). Initially, the application of a rather short anaerobic HRT (0.5 h, see Table 5.1) might be considered the main limiting factor for biological P-removal that led to the implementation of supplementary chemical dosing. If phosphorus is chemically removed it is not further available for PAO to replenish their intracellular poly-P pools, one of their main energy sources for anaerobic HAc uptake, inhibiting the biological phosphorus removal process. Nevertheless, an intermittent chemical dosing at Katwoude WWTP (applied when P effluent concentration exceeds 0.75 mg/L), to ensure a satisfactory P-removal, did not seem to cause any negative effect on EBPR activity (Table 5.3 and Figure 5.1). This may suggest that if supplementary chemical dosing is required an adequate implementation and use are necessary to prevent deterioration of the EBPR process. An adequate P-effluent set point to operate the supplementary chemical dosing is a key operating factor to avoid an inhibitory effect on EBPR.

The fractions of *Competibacter* were practically constant in all wastewater treatment plants but at lower fractions (1.7 ± 0.4 %) than reported elsewhere (Saunders *et al.*, 2003; Wong *et al.*, 2005; Tykesson *et al.*, 2006). The low sewerage temperature during sampling (12 ± 2 $^{\circ}$C) is likely the most important environmental factor since temperatures below 20 °C are not favourable for GAO (Whang and Park, 2006; Lopez-Vazquez *et al.*, 2007).

Defluviicoccus-related microorganisms Clusters 1 and 2 were hardly seen in any of the WWTP of the present study and *Sphingomonas* were not detected (Table 5.4). This is consistent with previous observations. *Defluviicoccus-related* organisms have been observed in low abundance in a limited number of full-scale EBPR systems (Wong *et al.*, 2004; Meyer *et al.*, 2006). Moreover, the WTTP from this survey treated mostly domestic wastewater whereas the occurrence of *Defluviicoccus* has been linked to industrial discharges (Burow *et al.*, 2007). At full-scale plants, *Sphingomonas* seem not to play a role as possible competitors of PAO since these microorganisms have not been observed in the current and previous studies (Meyer *et al.*, 2006; Burow *et al.*, 2007) and their presence appear to be restricted to lab-scale EBPR systems (Beer *et al.*, 2004; Wong *et al.*, 2007).

5.4.2. Environmental and operational parameters affecting the occurrence and activity of PAO and GAO

The influence of environmental and operational parameters on the occurrence and activity of PAO and GAO was determined based on the results of the statistical analysis (Table 5.5).

According to the observed significant correlations ($p < 0.05$), increases in the concentration of organic matter (as BOD and COD) in wastewater influents seemed to favour the growth of *Competibacter* (Table 5.5, correlations No. 6 - 9) although, according to effluent characteristics (Table 5.2), GAO fractions were not in the range that might have affected the EBPR process performance. This implies that an excess of carbon source tends to favour the development of GAO. Brdjanovic *et al.* (2000) reported that HAc overdosing at Haarlem Waarderpolder WWTP for biological P-removal purposes apparently resulted in the proliferation of GAO as indicated by the low measured P/HAc ratio (0.31 mgPO$_4^{3-}$-P/mgHAc, Table 5.3). During the present survey, HAc overdosing did not take place anymore at this plant leading to a higher P/HAc ratio (0.40 mgPO$_4^{3-}$-P/mgHAc, Table 5.3) and a low GAO fraction (0.4 ± 0.1 %, Table 5.4). Therefore, if external carbon source (such as VFA) should be added at full-scale systems, it is advised to carefully examine its application and limit its addition to an adequate (optimum) concentration.

Interestingly, no significant correlations were found between the occurrence and activity of *Accumulibacter* and *Competibacter* and VFA fractions (Table 5.5). HAc was the predominant VFA in the influents (0.83 ± 0.14 mg HAc/mg VFA) while HPr was also present but obviously comprising a lower

VFA fraction (0.13 ± 0.17 mg HPr/mg VFA). According to Oehmen *et al.* (2004, 2005), this should have been beneficial for *Competibacter*. However, *Accumulibacter* were the dominant microorganisms in all studied WWTPs (Figure 5.1). Rather than the VFA fractions other factors appear to have a higher influence on the PAO-GAO competition as also suggested elsewhere (Kong *et al.*, 2006).

Accumulibacter fractions measured at full-scale systems had a significant (positive) linear correlation ($p < 0.05$) with pH level measured in anaerobic tanks (Table 5.5, correlation No. 2) what is in accordance with lab-scale observations that stated that higher pH values are favourable for PAO metabolism and detrimental for GAO (Smolders *et al.*, 1994; Filipe *et al.*, 2001b; Schuler and Jenkins, 2002). Therefore, the operation of full-scale wastewater treatment plants at higher pH levels seemed to be a relevant factor for achieving and maintaining a reliable EBPR process operation. According to the correlation observed between total N_{INF} *vs.* pH anaerobic tank (Table 5.5, correlation No. 28), a well-defined denitrification stage appears to be linked to the generation of a higher mixed liquor pH value in the anaerobic stage resulting potentially favorable for *Accumulibacter*.

5.4.3. Occurrence of DPAO

A strong relationships ($p < 0.05$) between total total P_{EFF} *vs.* $NOx-N_{EFF}$ (Table 5.5, correlation No. 28) implied that either good biological P-removal is linked with good N-removal as consequence of well-operating conditions or that substantial fractions of PAO capable of denitrifying (DPAO) were present in the WWTPs of this survey. Despite that the former statement is supported by the observed high nutrient removal efficiencies, the latter seems to be likely based on the significant correlations among anoxic operating conditions (or related to nitrogen removal), *Accumulibacter* fractions and DPAO's activity (correlations No. 1, 11 − 12, 26 and 29: *Accumulibacter* fractions *vs.* total N_{INF}, Anoxic HRT *vs.* Aerobic P-uptake rate, Anoxic HRT *vs.* Anoxic P-uptake rate, total N_{INF} *vs.* HAc uptake rate or Aerobic P-uptake rate *vs.* Anoxic P-uptake rate). This may indicate that DPAO play an important role in nutrient removal at these full-scale systems.

In the present study, different Anoxic/Aerobic P-uptake rate ratios were measured (Table 5.3). In order to explain the different observed activities, Kerr-Jespersen and Henze (1993) and Meinhold *et al.*, (1999) proposed that divers Anoxic/Aerobic P-uptake rate ratios reflect the levels of enrichment of DPAO capable of using oxygen and nitrate and of non-denitrifying PAO only able to use oxygen. However, Kuba *et al.* (1996a, 1997b) and Wachtmeister *et al.* (1997), based on both lab-scale studies performed in an anaerobic-anoxic-aerobic HAc-fed reactor and full-scale studies, concluded that the Anoxic/Aerobic P-uptake rate ratio appeared to be a measure of the level of denitrifying capacity induced in PAO.

Kuba *et al.* (1996a) proposed that DPAO needed an exposure to NO_3^--N to stimulate the generation of the denitrifying biochemical pathway through the production of the required enzyme (nitrate reductase). In activated sludge plants without nitrogen removal (not exposed to anoxic conditions) the nitrate reductase enzyme is present in a limited amount leading to a low anoxic P-uptake activity (around 15 %). Meanwhile the cytochrome oxidation enzyme required for aerobic P-uptake is always present (Kuba *et al.*, 1996a). This hypothesis is in accordance with the correlations found in this study where the anoxic HRT was significantly correlated with the anoxic P-uptake activity (Table 5.5, correlation No. 12). Zeng *et al.* (2003a) also operated a similar laboratory set up like Kuba *et al.* (1996a) reporting that PAO and DPAO were the same microorganisms. This supports the hypothesis that the DPAO activity is induced through the exposition to anoxic environments.

Kuba *et al.* (1996a, 1997b) also proposed that only the implementation of predenitrification tanks (like in UCT plant configurations) can create the necessary conditions for the acclimation of DPAO. In the present study, higher Anoxic P-uptake activity was observed in plants operated with defined predenitrification stages (Hardenberg, Deventer, Katwoude, Hoek van Holland and Haarlem Waarderpolder WWTPs) than in plants where these tanks were not well-defined or absent (Venlo and Waarde WWTPs) (Tables 5.1 and 5.3) indicating that a good plant configuration is necessary for the development of DPAO. The low Anoxic/Aerobic P-uptake activities observed in this survey could be explained assuming that the longer (excessive) aeration periods, required during winter for the nitrification process, might have caused the oxidation of internal storage compounds (such as poly-β-hydroxybutyrate, PHB) (Brdjanovic *et al.*, 1998; Lopez *et al.*, 2006). Consequently, the intracellular PHB storage might have been (almost) depleted and not further used for denitrification or P-uptake (Kuba *et al.*, 1996a). This might have been the case of Venlo and Waarde WWTPs since the operation of their aeration systems depended on reaching certain low NH_4-N effluent concentrations (Table 5.1). Another hypothesis states that long aerobic periods might have caused a limited induction of denitrifying enzymes (Kuba *et al.*, 1996a).

Based on observed results and contrary to the significant contribution of DPAO, no significant correlations were found among the activity and occurrence of GAO and parameters related to denitrification. This might suggest that the plant operating conditions did not favour the development of DGAO. However, DPAO, that apparently have a similar metabolism like

DGAO, were present and seemed to be importantly involved in the nutrient removal. Allegedly, any possible denitrifying role by DGAO at the selected full-scale EBPR systems was not observed.

5.4.4. Effect of sewage temperature and involvement of other microbial communities

In the surveyed WWTPs, PAO comprised around 9 % of total biomass while GAO were present in relatively low fractions (2 %). This fact may explain the good EBPR process performance since, according to previous reports (Whang and Park, 2006; Lopez-Vazquez *et al.*, 2007), temperature lower than 20 ° C might have favoured PAO over GAO resulting beneficial for the stability of the EBPR process whereas the opposite would occur at higher temperature (> 20 °C).

Recently, the aerobic P-uptake, poly-P accumulation and abundance of *Actinobacteria* at full-scale EBPR systems (up to 29 % of total bacterial volume) suggest that these microorganisms may play a major role in the biological phosphorus removal process (Eschenhagen *et al.*, 2003; Wong *et al.*, 2005; Kong *et al.*, 2005, 2006). Unlike *Accumulibacter*, *Actinobacterial*-PAO are able to take up complex organic matter (like aminoacids) but the identity of the intracellular storage compound is still unknown (Kong *et al.*, 2005). Efforts should be undertaken at either lab- and full-scale systems not only to get a better insight about their metabolism but also to determine their possible contribution at full-scale biological P-removal processes.

A similar study under summer conditions (sewage temperature higher than 20 °C) executing lab-activity tests with acetate and (synthetic) wastewater influent rich in aminoacids (such as casaminoacids) may be helpful to assess, confirm or deny the observations drawn from the present and previous studies related to the involvement of different microbial communities at full-scale EBPR systems. This could lead to the design and operation of better and more reliable EBPR WWTP.

5.5. Conclusions

The influence of operating and environmental conditions on the microbial communities of the EBPR process at Dutch full-scale activated sludge WWTP under winter conditions (sewage temperature around 12 °C) was studied. Well-defined and operated denitrification stages are important to favour the growth of PAO (*Accumulibacter*) and stimulate the development of Denitrifying PAO (DPAO) influencing positively the nitrogen and phosphorus removal process. The denitrifying capacity of DPAO appeared to be inducible through the exposition to anoxic environments. pH values were positively correlated with the occurrence of *Accumulibacter*. A positive correlation was found between the occurrence of *Competibacter* and organic matter concentrations (as BOD and COD) present in the influent. However, *Competibacter* did not cause a major effect on the EBPR process performance because the observed fractions were not in the range that would have led to biological phosphorus removal deterioration. Likely, the low average sewerage temperature (12 ± 2 °C) limited their proliferation. *Defluviicoccus*-related microorganisms were only observed in negligible fractions (< 0.1 %) in a few WWTP from the present survey. Meanwhile, *Sphingomonas* were not seen in any plant.

Acknowledgements

The authors would like to acknowledge the National Council for Science and Technology from Mexico (CONACYT) and the Autonomous University of the State of Mexico for the grant awarded to Carlos Manuel Lopez Vazquez. The valuable help and support provided by Jos van der Ent and Joost van den Bulk (BSc. trainees) is highly appreciated. Special and sincere thanks to all water boards and practitioners who kindly collaborated in this research: H. van der Spoel (Katwoude, Uitwaterende Sluizen in Hollands Noorderkwartier), A. de Man (Venlo, Limburg), H. van Veldhuizen (Deventer, Groot Salland), M. Augustijn (Waarde, Zeeuwse Eilanden), J. W. Koelewijn (Amersfoort, Vallei en Eem), E. Majoor, (Hardenberg, Velt & Vecht), A. Vermaat (Haarlem Waarderpolder, Rijnland) and E.G. Schuurman (Hoek van Holland, Delftland).

References

APHA/AWWA (1995) Standard methods for the examination of water and wastewater, 19th ed. Port City Press, Baltimore.

Amann RI (1995) In situ identification of microorganisms by whole cell hybridization with rRNA-targeted nucleic acid probes. In: Akkermans, A.D.L.; van Elsas, J.D.; de Bruijn, F.J. (Eds.), Molecular Microbial Ecology Manual. Kluwer Academic Publications, Dordrecht, The Netherlands.

Amman RI, Ludwig W, Schleifer KH (1995). Philogenetic identification and *in situ* detection of individual microbial cells without cultivation. Microbiol Revs. 59, 143-169.

Beer M, Kong YH, Seviour RJ (2004) Are some putative glycogen accumulating organisms (GAO) in anaerobic: aerobic activated sludge systems members of the *α-Proteobacteria*? Microbiology-SGM. 150, 2267–2275.

Brdjanovic D, van Loosdrecht MCM, Versteeg P, Hooijmans CM, Alaerts GJ, Heijnen JJ (2000) Modeling COD, N and P removal in a full-scale wwtp Haarlem Waarderpolder. Water Res. 34, 846–858.

Brdjanovic D, Slamet A, van Loosdrecht MCM, Hooijmans CM, Alaerts GJ, Heijnen JJ (1998) Impact of excessive aeration on biological phosphorus removal from wastewater. Water Res. 32, 200–208.

Burow LC, Kong Y, Nielsen JL, Blackall LL, Nielsen PH (2007) Abundance and ecophysiology of Defluviicoccus spp., glycogen-accumulating organisms in full-scale wastewater treatment processes. Microbiology-SGM 153, 178-185.

Cech JS, Hartman P (1993) Competition between phosphate and polysaccharide accumulating bacteria in enhanced biological phosphorus removal systems. Water Res. 27, 1219–1225.

Chua ASM, Onuki M, Satoh H, Mino T (2006) Examining substrate uptake patterns of *Rhodocyclus*-related PAO in full-scale EBPR plants by using the MAR-FISH technique. Water Sci. Technol. 54, 63–70.

Crocetti GR, Banfield JF, Keller J, Bond PL, Blackall LL (2002) Glycogen accumulating organisms in laboratory-scale and full-scale wastewater treatment processes. Microbiology–SGM 148, 3353–3364.

Crocetti GR, Hugenholtz P, Bond PL, Schuler A, Keller J, Jenkins D, Blackall LL, (2000) Identification of polyphosphate-accumulating organisms and design of 16S rRNA-directed probes for their detection and quantitation. Appl. Environ. Microbiol. 66, 1175–1182.

Daims H, Bruhl A, Amman R, Schleifer K-H, Wagner M (1999) The domain-specific probe EUB 338 is insufficient for the detection of all bacteria: development and evaluation of a more comprehensive probe set. Syst. Appl. Microbiol. 22, 434 - 444.

Eschenhagen M, Schuppler M, Roske I (2003) Molecular characterization of the microbial community structure in two activated sludge systems for the advanced treatment of domestic effluents. Water Res. 37, 3224 - 3232.

Filipe CDM, Daigger GT, Grady CPL (2001a) A metabolic model for acetate uptake under anaerobic conditions by glycogen accumulating organisms: stoichiometry, kinetics and effect of pH. Biotechnol. Bioeng. 76, 17–31.

Filipe CDM, Daigger GT, Grady CPL (2001b) pH as a key factor in the competition between glycogen-accumulating organisms and phosphorus-accumulating organisms. Water Environ.. Res. 73, 223–232.

Hu ZR, Wentzel MC, Ekama GA (2002) Anoxic growth of phosphate-accumulating organisms (PAOs) in biological nutrient removal activated sludge systems. Water Res. 36, 4927–4937.

Kerrn-Jespersen JP, Henze M (1993) Biological phosphorus uptake under anoxic and aerobic conditions. Water Res. 27, 617–624.

Kong YH, Ong SL, Ng WJ, Liu WT (2002) Diversity and distribution of a deeply branched novel proteobacterial group found in anaerobic–aerobic activated sludge processes. Environ. Microbiol. 4, 753–757.

Kong Y, Nielsen JL, Nielsen PH (2004) Microautoradiography study of *Rhodocyclus*-related polyphosphate accumulating bacteria in full-scale enhanced biological phosphorus removal plants. Appl. Environ. Microbiol. 70, 5383–5390.

Kong Y, Nielsen JL, Nielsen PH (2005) Identity and ecophysiology of uncultured actinobacterial polyphosphate-accumulating organisms in full-scale enhanced biological phosphorus removal plants. Appl. Environ. Microbiol. 71, 4076 - 4085.

Kong YH, Xia Y, Nielsen JL, Nielsen PH (2006) Ecophysiology of a group of uncultured Gammaproteobacterial glycogen accumulating organisms in full-scale enhanced biological phosphorus removal wastewater treatment plants. Environ. Microbiol. 8, 479–489.

Kuba T, van Loosdrecht MCM, Heijnen JJ (1996a) Effect of cyclic oxygen exposure on the activity of denitrifying phosphorus removing bacteria. Water Sci. Technol. 34, 33–40.

Kuba T, van Loosdrecht MCM, Heijnen JJ (1996b) Phosphorus and nitrogen removal with minimal COD requirement by integration of denitrifying dephosphatation and nitrification in a two sludge system. Water Res. 30, 1702–1710.

Kuba T, van Loosdrecht MCM, Heijnen JJ (1997a) Biological dephosphatation by activated sludge under denitrifying conditions: pH influence and occurrence of denitrifying dephosphatation in a full-scale wastewater treatment plant. Water Sci. Technol. 36, 75–82.

Kuba T, van Loosdrecht MCM, Brandse FA, Heijnen JJ (1997b) Occurrence of denitrifying phosphorus removing bacteria in modified UCT-type wastewater treatment plants. Water Res. 31, 777–786.

Liu WT, Nakamura K, Matsuo T, Mino T (1997) Internal energy-based competition between polyphosphate- and glycogen-accumulating bacteria in biological phosphorus removal reactors-effect of P/C feeding ratio. Water Res. 31, 1430–1438.

Lopez C, Pons MN, Morgenroth E (2006) Endogenous process during long-term starvation in activated sludge performing enhanced biological phosphorus removal. Water Res. 40, 1519–1530.

Lopez-Vazquez CM, Song YI, Hooijmans CM, Brdjanovic D, Moussa MS, Gijzen HJ, van Loosdrecht MCM (2007) Short-term temperature effects on the anaerobic metabolism of glycogen accumulating organisms. Biotech. Bioeng. 97, 483–495.

Meinhold J, Filipe CDM, Daigger GT, Isaacs S (1999) Characterization of the denitrifying fraction of phosphate accumulating organisms in biological phosphorus removal. Water Sci. Technol. 39, 31–42.

Meyer RL, Saunders AM, Blackall LL (2006) Putative glycogen accumulating organisms belonging to *Alphaproteobacteria* identified through rRNA-based stable isotope probing. Microbiology-SGM. 152, 419–429.

Oehmen A, Yuan Z, Blackall LL, Keller J (2004) Short-term effects of carbon source on the competition of polyphosphate accumulating organisms and glycogen accumulating organisms.Water Sci. Technol. 50, 139–144.

Oehmen A, Yuan Z, Blackall LL, Keller J (2005) Comparison of acetate and propionate uptake by polyphosphate accumulating organisms and glycogen accumulating organisms. Biotechnol. Bioeng. 91, 162–168.

Satoh H, Mino T, Matsuo T (1994) Deterioration of enhanced biological phosphorus removal by the domination of microorganisms without polyphosphate accumulation. Water Sci. Technol. 30, 203–211.

Saunders AM, Oehmen A, Blackall LL, Yuan Z, Keller J (2003) The effect of GAOs (glycogen accumulating organisms) on anaerobic carbon requirements in full-scale Australian EBPR (enhanced biological phosphorus removal) plants. Water Sci. Technol. 47, 37–43.

Schuler AJ, Jenkins D (2003) Enhanced biological phosphorus removal from wastewater by biomass with different phosphorus contents, Part 1:

Experimental results and comparison with metabolic models. Water Environ. Res. 75, 485–498.

Schuler AJ, Jenkins D (2002) Effects of pH on enhanced biological phosphorus removal metabolisms. Water Sci. Technol. 46, 171–178.

Seviour RJ, Mino T, Onuki M (2003) The microbiology of biological phosphorus removal in activated sludge systems. FEMS Microbiol. Reviews 27, 99–127.

Smolders GJF, van Der Meij J, van Loosdrecht MCM, Heijnen JJ (1994) Model of the anaerobic metabolism of the biological phosphorus removal processes: stoichiometry and pH influence. Biotechnol. Bioeng. 43, 461–470.

Smolders GJF, van Loosdrecht MCM, Heijnen JJ (1996) Steady-state analysis to evaluate the phosphate removal capacity and acetate requirement of biological phosphorus removing mainstreams and sidestream process configurations. Water Res. 30, 2748–2760.

Sudiana IM, Mino T, Satoh H, Nakamura K, Matsuo T (1999) Metabolism of enhanced biological phosphorus removal and non-enhanced biological phosphorus removal sludge with acetate and glucose as carbon source. Water Sci. Technol. 39, 29–35.

Thomas M, Wright P, Blackall LL, Urbain V, Keller J (2003) Optimization of Noosa BNR plant to improve performance and reduce operating costs. Water Sci. Technol. 47, 141–148.

Tykesson E, Blackall LL, Kong Y, Nielsen PH, Jansen JC (2006) Applicability of experience from laboratory reactors with biological phosphorus removal in full-scale plants. Wat Sci Technol. 54, 267–275.

van Loosdrecht MCM, Brandse FA, Vries AC (1998) Upgrading of wastewater treatment processes for integrated nutrient removal – The BCFS process. Water Sci. Technol. 37, 209–217.

Wachtmeister A, Kuba T, van Loosdrecht MCM, Heijnen JJ (1997) A sludge characterization assay for aerobic and denitrifying phosphorus removing sludge. Water Res. 31, 471–478.

Whang LM, Park JK (2006) Competition between polyphosphate- and glycogen-accumulating organisms in enhanced-biological-phosphorus-removal systems: effect of temperature and sludge age. Water Environ. Res. 78: 4–11.

Wong MT, Tan FM, Ng WJ, Liu WT (2004) Identification and occurrence of tetrad-forming Alphaproteobacteria in anaerobic–aerobic activated sludge processes. Microbiology-SGM. 150, 3741–3748.

Wong MT, Mino T, Seviour RJ, Onuki M, Liu WT (2005) In situ identification and characterization of the microbial community structure of full-scale enhanced biological phosphorus removal plants in Japan. Water Res. 39, 2901–2914.

Wong MT, Liu WT (2007) Ecophysiology of *Defluviicoccus*–related tetra-forming organisms in an anaerobic-aerobic activated sludge process. Environ Microbiol. 9, 1485–1496.

Zeng RJ, Saunders AM, Yuan Z, Blackall LL, Keller J (2003a) Identification and comparison of aerobic and denitrifying polyphosphate-accumulating organisms. Biotechnol Bioeng. 83, 140–148.

Zeng RJ, Yuan Z, Keller J (2003b) Enrichment of denitrifying glycogen-accumulating organisms in anaerobic/anoxic activated sludge system. Biotechnol. Bioeng. 81, 397–404.

Zilles JL, Peccia J, Kim MW, Hung CH, Noguera D (2002) Involvement of *Rhodocyclus*-related organisms in phosphorus removal in full-scale wastewater treatment plants. Appl. Environ. Microbiol. 68, 2763–2769.

6

Chapter

A practical method for the quantification of PAO and GAO populations in full-scale systems

Content

This chapter has been published as:
    Lopez-Vazquez CM, Hooijmans CM, Brdjanovic D, Gijzen HJ, van Loosdrecht MCM (2007) A practical method for the quantification of phosphorus- and glycogen-accumulating organisms populations in activated sludge systems. *Water Environ Res* 79(13):2487-2498.

Abstract

A simple method for the quantification of PAO and GAO population fractions in activated sludge systems is presented. In order to develop such a method the activity observed in anaerobic batch tests executed with different PAO-GAO ratios by mixing highly enriched PAO and GAO cultures was studied. Strong correlations between PAO-GAO population ratios and biomass activity were observed ($R^2 >$ 0.97). This served as a basis for the proposal of a simple and practical method to quantify the PAO and GAO populations in activated sludge systems based on commonly measured and reliable analytical parameters (such as mixed liquor suspended solids, acetate and orthophosphate) without requiring molecular techniques. This method relies on the estimation of the total active biomass population under anaerobic conditions (PAO plus GAO populations) by measuring the maximum acetate uptake rate in the presence of excess acetate. Later on, the PAO and GAO populations present in the activated sludge system can be estimated by taking into account the PAO-GAO ratio calculated on the basis of the anaerobic P-release/HAc consumed ratio. The proposed method was evaluated using activated sludge from municipal wastewater treatment plants. The results from the quantification performed following the proposed method were compared to direct population estimations carried out with FISH analysis (determining *Candidatus Accumulibacter Phosphatis* as PAO and *Candidatus Competibacter Phosphatis* as GAO). The method showed to be potentially suitable to estimate the PAO and GAO populations regarding the total PAO-GAO biomass. It could be used not only to evaluate the performance of EBPR systems but also in the calibration of potential activated sludge mathematical models regarding the PAO-GAO coexistence.

6.1. Introduction

From both EBPR process evaluation and modeling perspectives the estimation of PAO and GAO populations in activated sludge systems is a relevant issue. Currently, PAO and GAO population contribution to total biomass can only be satisfactorily estimated by applying molecular techniques (e. g. Fluorescence *in situ* Hybridization, FISH, Amann, 1995; and, Fluorescence *in situ* Hybridization - Microautoradiography, FISH-MAR, Lee *et al.*, 1999). These techniques require sophisticated and expensive equipment that significantly restrict its wide use.

Zeng *et al.* (2003b) and Yagci *et al.* (2004) proposed two different combined metabolic models to study the anaerobic metabolism of a PAO-GAO mixed culture. These models were able to satisfactorily describe and predict the activity of a mixed PAO-GAO culture and estimate the PAO and GAO populations. However, they were determined based on lab-scale experiments and relied on the determination of PHA fractions. This fact might limit their applicability considering that storage compounds (such as PHA and glycogen) comprise only a small part of volatile suspended solids in full-scale activated sludge systems due to the low PAO and GAO active biomass fractions, making PHA and glycogen difficult to determine. Furthermore, those proposed models were only validated at a defined PAO-GAO ratio possibly requiring further research to validate them and, if needed, extend them.

Brdjanovic *et al.* (1999) proposed a bioassay to estimate the GAO/PAO ratio in activated sludge. This method was based on acetate and orthophosphate measurements with and without depletion of the poly-P pool. Nevertheless, the method was not experimentally validated on activated sludge. Later, Filipe *et al.* (2001b) and Schuler and Jenkins (2003) proposed other simple methods to estimate the PAO and GAO dominance based on the observed EBPR activity. Filipe *et al.* (2001b) proposed to evaluate the possible presence of GAO by executing two anaerobic batch tests at different pH-values (one at pH 6.5 and other at pH 8.0) and measuring the glycogen consumed and PHA accumulated. Schuler and Jenkins (2003) suggested using either the anaerobic P-released/acetate consumption ratio or the glycogen degradation/acetate consumption ratio. However, these assays provide a means of determining the relative PAO and GAO populations or activities, but they do not by themselves provide an estimate of the PAO and GAO fractions of the total biomass. Moreover, the applicability of these methods could be limited as PHA and glycogen are parameters difficult to determine at full-scale activated sludge wastewater treatment plants.

The aim of this study is to develop a simple and practical method to quantify the PAO and GAO populations at lab- and full-scale systems as hypothesized by Brdjanovic *et al.* (1999). The method is based on commonly measured and reliable analytical parameters, such as acetate (HAc) and orthophosphate (PO_4^{3-}-P), without requiring molecular methods like FISH. The proposed method was developed using synthetic mixtures of (highly) enriched cultures of PAO and GAO and validated with samples from full-scale systems.

6.2. Materials and methods

6.2.1. Enrichment of PAO and GAO cultures

PAO and GAO cultures were enriched in two separate arguably identical double-jacketed lab-scale reactors at 20 ± 0.5 °C. Reactors were operated and controlled automatically in a sequential mode (SBR). Each SBR had a working volume of 2.5 L. Activated sludge from a municipal wastewater treatment plant with a 5-stage Bardenpho configuration (Hoek van Holland in The Netherlands) was used as inoculum. SBRs were operated in cycles of 6 hours (2.25 h anaerobic, 2.25 aerobic and 1.5 settling phase) following similar operating conditions used in previous studies (Smolders *et al.*, 1994a; Brdjanovic *et al.*, 1997). Further details regarding the operation of the reactors can be found in chapter 2 and 3.

The main difference between the operating conditions of PAO and GAO SBR (called SBR1 and SBR2, respectively) was the phosphorus (P) content in the synthetic medium supplied to each reactor. Influent of SBR1 contained 15 mg PO_4^{3-}-P/L (0.48 P-mmol/L) (Smolders *et al.*, 1994a) while P content in GAO SBR influent was limited to 2.2 mg PO_4^{3-}-P/L (0.07 P-mmol/L) (Liu *et al.*, 1997). Besides the different P concentrations, synthetic media contained per litre: 850 mg NaAc·3H$_2$O (12.5 C-mmol/L, approximately 400 mg COD/L) and 107 mg NH$_4$Cl (2 N-mmol/L). In addition, 2 mg of allyl-N thiourea (ATU) were added to inhibit nitrification. The rest of minerals and trace metals present in the synthetic media were prepared as described by Smolders *et al.*, (1994a). Prior to use, synthetic media were autoclaved at 110 °C for 1 h.

The performance of SBR1 and SBR2 was regularly monitored by measuring orthophosphate (PO_4^{3-}-P), mixed liquor suspended solids (MLSS) and mixed liquor volatile suspended solids (MLVSS). Cycle measurements were carried out to determine the biomass activity when both SBR reached steady-state conditions. Additionally, quantification of bacterial populations was undertaken via FISH analysis to define the enrichment level of PAO and GAO in each SBR.

Anaerobic batch experiments were performed in a separate (third) batch reactor after SBR1 and 2 reached steady-state operation. This means that SBR1 and 2 were only used as sources of enriched cultures to perform the anaerobic batch tests at different PAO/GAO mixing ratios.

6.2.2. Operation of batch reactor

A double-jacketed laboratory fermenter with a maximal volume of 0.5 L was used for the execution of anaerobic batch experiments with the different PAO/GAO ratios. Experiments were performed at controlled temperature and pH (20 ± 0.5 °C and 7.0 ± 0.05, respectively). pH was automatically maintained by dosing of 0.2 M HCl and 0.2 M NaOH. At the beginning of each experiment, enriched PAO and/or GAO sludge (depending on the corresponding experiment) were manually transferred from respective SBR to batch reactor. By transferring a maximum amount of 250 mL of sludge (which is the amount of sludge daily wasted from each SBR) disturbance of continuous operation and steady-state conditions of SBR 1 and 2 was minimized. Sludge used in batch tests was not returned to SBRs. N$_2$ gas was continuously introduced to the reactor during the entire duration of the test at a flow rate of 30 L/h. During the anaerobic batch experiments the sludge was constantly stirred at 500 rpm.

6.2.3. Batch tests with different PAO/GAO population ratios

At the end of the aerobic phase, a defined volume of enriched PAO and/or GAO cultures was transferred from SBR1 and/or SBR2 to batch reactor. Volume was determined according to the proposed PAO/GAO population ratios based on average biomass concentrations (as volatile suspended solids) measured in PAO and GAO SBRs under steady-state conditions. Six different PAO/GAO mixing ratios were studied: 100/0, 80/20, 60/40, 40/60, 20/80 and 0/100, as PAO/GAO percentages. After sludge transfer, N_2 gas was supplied to create and maintain anaerobic conditions in the batch reactor. Following the first 5 minutes, concentrated synthetic medium was added in excess as pulse. The initial acetate (HAc) concentration in the reactor at the start of the batch test was determined based on preliminary batch tests performed only with PAO and GAO enriched biomass. The aim was to add HAc in excess to avoid substrate-limiting conditions during execution of the tests but avoiding acetate inhibition (toxicity). After substrate addition, biomass activity was continuously followed during 5 h. N_2 gas was supplied during the entire experiment to avoid oxygen intrusion.

HAc uptake rate and P-release rate were the kinetic parameters of interest in this study. Specific rates were calculated considering the HAc consumption or P-release profiles and active biomass concentrations for each anaerobic batch test. Active biomass concentration was determined as MLVSS but excluding PHB, PHV and glycogen contents (Active Biomass = MLVSS – PHB – PHV – glycogen). Active biomass was used in order to distinguish between biomass and organic and inorganic storage products since internal storage products can make up half of the biomass (either PAO or GAO). Besides, PHB, PHV and glycogen fractions change strongly due to dynamics. To calculate the specific rates, the active biomass concentration was expressed in C-mol units by taking into account the experimentally determined composition of PAO ($CH_{2.09}O_{0.54}N_{0.20}P_{0.015}$) (Smolders *et al.* (1994b) and GAO ($CH_{1.84}O_{0.5}N_{0.19}$) (Zeng *et al.* (2003a).

Glycogen hydrolysis and PHB and PHV production per HAc consumed (Gly/HAc, PHB/HAc and PHV/HAc ratios, respectively) were the stoichoimetric parameters evaluated. Implying that a characterization of both biomass and liquid phase was performed during batch tests. Samples were collected for the determination of HAc, PO_4^{3-}-P, PHB, PHV, glycogen, MLSS and MLVSS.

6.2.4. Validation of the method

A method to quantify the PAO and GAO populations was proposed based on relationships and correlations observed between biomass activity from anaerobic batch tests and actual PAO and GAO fractions. In order to validate the method, anaerobic batch tests were performed with mixed liquor activated sludge from six different municipal wastewater treatment plants. Tests were performed with fresh activated sludge (< 24 h old) collected at the end of the aerobic tank. Batch tests were carried out applying identical operating conditions as in the lab-tests executed to develop the method (20 ± 0.5 °C and pH 7.0 ± 0.05). A double-jacketed laboratory fermentor with a maximal volume of 1.5 L was used for the execution of the anaerobic batch experiments. Before tests started, 20 mg of allyl-N-thourea were added to avoid nitrification during acclimation conditions and avoid the presence of nitrate or nitrite during the execution of batch experiments. 1 L of mixed liquor was continuously aerated for 1 h to acclimatize it to lab conditions and remove possible readily biodegradable organic matter that might be present in the sludge. Following the acclimation period, N_2 gas was introduced. After 10 minutes, test started with the addition of synthetic medium in a pulse mode. Initial HAc concentration in batch tests was 3.3 C-mmol/L (100 mg HAc/L, approximately 107 mg COD/L). N_2 gas was continuously introduced to the reactor during the whole test at a flow rate of 30 L/h to avoid oxygen intrusion. Sludge was stirred at 500 rpm. HAc consumption and PO_4^{3-}-P determination profiles were determined based on samples collected during the 5 h batch test length. Samples were collected every 5 min in the first 30 min of the batch tests and every 30 min in the remaining 4.5 h. Maximum acetate uptake and P-release rates were calculated based on the acetate and orthophosphate profiles observed in the first 30 min of the tests. Meanwhile, the anaerobic P-released/HAc ratios and total P-released were calculated considering the concentrations observed at the start and end of the tests. Quantifications obtained with this method were compared with PAO and GAO population fractions quantified via FISH analysis.

6.2.5. Analyses

PO_4^{3-}-P was determined by the ascorbic acid method. Analyses, including MLSS and MLVSS determination, were performed in accordance with Standard Methods (A.P.H.A., 1995). Meanwhile, NO_3^--N determination was performed with a Dionex ICS-1000 Ion Chromatographer (IC) (Dionex, The Netherlands). HAc was determined with a HP 5890 Gas Chromatographer (GC) equipped with an CB-FFAP column of 25 m x 0.53 mm x 1 mm. PHA content (as PHB and PHV) of freeze dried biomass was determined according to the method described by Smolders *et al.* (1994a). Glycogen determination was also executed according to the method described by

Smolders *et al.* (1994a) but extending the digestion phase at 100°C from 1 to 5 h.

For the quantification of the PAO and GAO communities (as *Candidatus Accumulibacter Phosphatis* and *Candidatus Competibacter Phosphatis*, respectively), Fluorescence *in situ* Hybridization (FISH) was performed as described in Amman (1995). The EUBMIX probe (mixture of probes EUB 338, EUB338-II and EUB338-III) to target the entire bacterial population, PAOMIX probe (mixture of probes PAO462, PAO651 and PAO 846) to target *Accumulibacter* (Crocetti *et al.*, 2000) and GAOMIX probe (mixture of probes GAOQ431 and GAOQ989) to target *Competibacter* (Crocetti *et al.*, 2002) were used to determine the PAO and GAO populations. FISH preparations were visualized with a Zeiss Axioplan 2 microscope.

Quantification was performed using the MATLAB image processing toolbox (The Mathworks, Natick, MA, US). 8-bit images for each of color channels (green for PAO, FLUOS; red for GAO, Cy3; and blue for EUB, Cy5) were converted into binary format using direct thresholding at a graylevel determined using the Otsu method (Otsu, 1979). Image coverage was computed by dividing the number of pixels corresponding to the object with the total number of pixels of the image. Fractions of PAO (*Accumulibacter*) and GAO (*Competibacter*) were calculated as the ratio between their image coverage and that of the entire bacterial population (EUBMIX). Around thirty separate images were evaluated per sample with final results reflecting the average fractions of PAO and GAO, expressed as percentage of EUB. The standard error of the mean (SEM) was calculated as the standard deviation of the area percentages divided by the square root of the number of images analysed.

6.3. Results

6.3.1. Enriched PAO and GAO cultures

SBR1 and SBR2 were operated for more than 80 days before the PAO/GAO mixing tests started. Both reactors reached steady-state cyclic conditions within the first 50 days of operation. Figure 6.1a presents a cycle measurement of SBR1 under steady-state conditions. It shows the typical anaerobic-aerobic cycle profile of an enriched PAO culture: complete anaerobic acetate consumption, substantial anaerobic P-release (anaerobic P/HAc ratio around 0.47 P-mol/HAc C-mol) and almost complete aerobic P-uptake (PO_4^{3-}-P concentration in effluent < 1 mg/L). In the anaerobic phase these conversions were associated with glycogen consumption and PHA production (as PHB and PHV) while, under aerobic conditions, they were coupled with PHA oxidation and glycogen production. Anaerobic P/HAc ratio was in the range of previous observed ratios for PAO cultures enriched at 20 °C and pH 7.0 (0.46 - 0.50 P-mol/HAc C-mol; Smolders *et al.*, 1994a; Kuba *et al.*, 1994; Oehmen *et al.*, 2005). At the end of the aerobic phase, MLSS and MLVSS were 3297 and 2256 mg/L, respectively, resulting in an average MLVSS/MLSS ratio of 0.68. This low MLVSS/MLSS ratio indicated high inorganic matter storage, presumably poly-P. High ash content (based on the low MLVSS/MLSS ratio) and the observed activity suggested that PAO were the dominant microorganisms in SBR1.

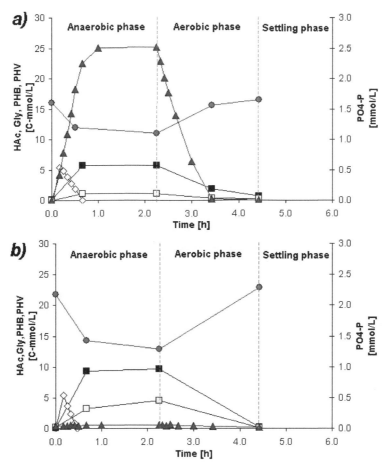

Figure 6.1. Cycle profiles observed in (a) an enriched PAO culture in SBR1 and (b) an enriched GAO culture in SBR2: acetate (open diamonds), glycogen (red circles), PHB (black squares), PHV (yellow squares) and orthophosphate (blue triangles).

On the other hand, the activity observed in SBR2 under steady-state conditions (Figure 6.1b), was characteristic of an enriched GAO culture: acetate fully consumed in the anaerobic phase associated with a negligible anaerobic P/HAc ratio (< 0.01 P-mol/HAc C-mol), glycogen consumption and PHA production. MLSS (2455 mg/L) and MLVSS (2366 mg/L) concentrations observed at the end of the aerobic phase led to a high MLVSS/MLSS ratio of 0.96 that indicated low inorganic matter storage.

Since nitrification was inhibited due to ATU addition, the slight consumption of ammonium observed in both reactors was associated to biomass growth requirements (data not shown).

The activity observed in both reactors suggested that PAO and GAO were sufficiently enriched in SBR1 and SBR2, respectively, for the purpose of this study. FISH analysis corroborated that PAO were the dominant microorganisms in SBR1 (Figure 6.2a) and GAO in SBR2 (Figure 6.2b). Quantification of PAO and GAO populations from SBR1 indicated that PAO, GAO and other bacteria contributed to total biomass in 93 ± 1 %, 4 ± 1 % and 3 ± 1 %, respectively. In SBR2, GAO comprised 92 ± 3 %, PAO less than 1 % and other bacteria 8 ± 3 % of the biomass present in SBR2. These results confirmed that highly enriched cultures of PAO and GAO were cultivated in each reactor and that the applied FISH probes seemed to be sufficiently specific to target the present PAO and GAO cultures. It also means that apparently there were no other significant populations in the reactors besides *Accumulibacter* and *Competibacter*.

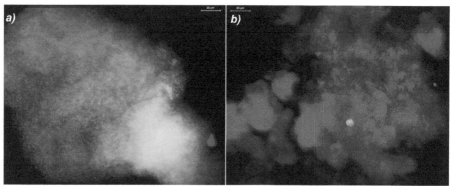

Figure 6.2. Bacterial population distribution by applying FISH techniques showing (a) an enriched PAO culture in SBR1 and (b) an enriched GAO culture in SBR2 (bar indicates 20 μm; blue: Eubacteria; green: PAO; red: GAO). PAO appear light green while GAO pink due to the superposition of EUB and PAO and GAO mix probes, respectively.

6.3.2. Anaerobic PAO/GAO mixing ratios tests

Figure 6.3 displays the biomass specific acetate consumption as measured in the six different PAO/GAO tests, in C-mol HAc/C-mol active biomass units. A suitable initial acetate concentration of 12.5 C-mmol/L was found via preliminary tests (375 mg HAc/L, approximately 400 mg COD/L). This initial concentration was determined through anaerobic batch tests executed only with PAO or GAO enriched biomass where different initial HAc concentrations were tested. HAc concentrations higher than 13.3 C-mmol/L (400 mg/L) caused a partial inhibition effect on GAO. When initial HAc concentration was $\geq$ 13.3 C-mmol/L, GAO were not able to immediately start to take up the substrate (data not shown). GAO needed about 20 minutes before starting utilizing HAc. On the other hand, PAO did not exhibit any inhibitory effect as GAO. The constant HAc uptake in the different anaerobic PAO/GAO ratios tests showed that selected initial concentration (12.5 C-mmol/L) did not cause any inhibitory effect (Figure 6.3). Independently of the different PAO/GAO mixing ratios, similar maximum HAc uptake rates were observed (0.20 – 0.24 HAc C-mol/C-mol active biomass/h). This shows that, under the applied operating conditions (20 $^{\circ}$C, pH 7.0 $\pm$ 0.05), the competition between PAO and GAO was very stringent due to their similar acetate uptake rates. Besides, substrate seemed not to be limiting, there was always at least a small amount of HAc remaining at the end of the test (around 0.2 C-mmol HAc/L). However, major differences are observed between the first and the second hour of the tests when inflexion points in profiles were found.

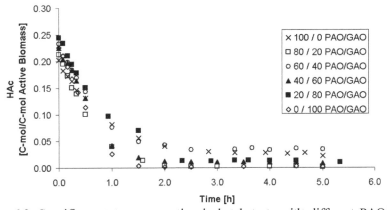

Figure 6.3. Specific acetate consumption in batch tests with different PAO/GAO ratios: 0/100 ($\Diamond$), 20/80 ($\blacksquare$), 40/60 ($\blacktriangle$), 60/40 ($\circ$), 80/20 ($\square$) and 100/0 ($\times$).

Figure 6.4 shows the specific P-release observed in the six different PAO/GAO tests, (expressed as P-mol/C-mol active biomass). As expected, a clear trend was observed where P-release profiles increased with the PAO fraction in the biomass.

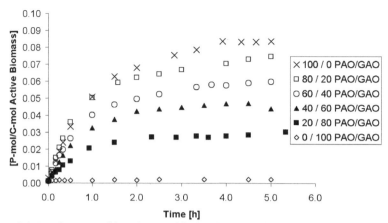

Figure 6.4. P-release profiles observed in batch tests with different PAO/GAO ratios: 0/100 (◊), 20/80 (■), 40/60 (▲), 60/40 (○), 80/20 (□) and 100/0 (×).

6.3.3. Kinetics of the PAO/GAO mixing ratios tests

The intrinsic kinetics observed in the six different PAO/GAO batch tests were expressed in terms of actual PAO and GAO fractions. These fractions were calculated based on the estimated PAO and GAO biomass transferred from parent SBR to parallel batch reactor and corrected according to the biomass quantification performed via FISH analyses. Measured HAc uptake rates, anaerobic P/HAc ratios, specific P-release rates and total P-released per active biomass relationships are displayed in Figure 6.5.

Similar HAc uptake rates (ranging from 0.16 to 0.23 C-mol/C-mol active biomass/h) were observed in the six different anaerobic batch tests (Figure 6.5a) suggesting that PAO-GAO competition took place in a narrow kinetic range. In particular, HAc uptake rates carried out at 0/100 and 100/0 PAO/GAO ratios (enriched cultures) were practically identical (0.19 and 0.20 C-mol/C-mol active biomass/h, respectively). Table 6.1 shows that measured rates were in the range of previous reported values under similar operating conditions (7.0 pH, 20 °C).

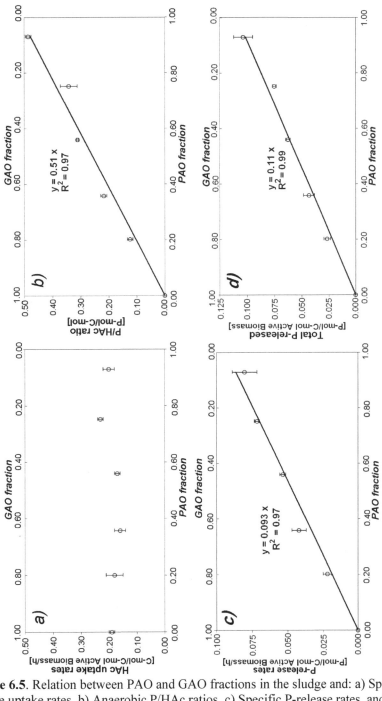

Figure 6.5. Relation between PAO and GAO fractions in the sludge and: a) Specific acetate uptake rates, b) Anaerobic P/HAc ratios, c) Specific P-release rates, and d) Total P-released per active biomass. PAO and GAO fractions were determined by FISH.

Table 6.1. Comparison of maximum acetate uptake rates of PAO and GAO with previous studies at pH 7.0 and 20 °C.

Microorganism	Acetate uptake rate [C-mol/C-mol active biomass/h]	Reference
PAO	0.27	Smolders *et al.* (1994a)
	0.20	Filipe *et al.* (2001b)
	0.20	Kuba *et al.* (1996)
	0.20	Brdjanovic *et al.* (1997)
	0.20 ± 0.03	Lopez-Vazquez *et al.* (2007)
	0.20 ± 0.02	This study
GAO	0.24	Filipe *et al.* (2001a)
	0.16 – 0.18	Zeng *et al.* (2003a)
	0.20 ± 0.01	Lopez-Vazquez *et al.* (2007)
	0.19 ± 0.01	This study

A strong correlation ($R^2 = 0.97$) was observed when anaerobic P/HAc ratios measured in the batch tests were plotted *versus* actual PAO and GAO fractions (Figure 6.5b). The highest P/HAc ratio (0.47 P-mol/HAc C-mol) was found at the 100/0 test when PAO comprised approximately 93 % of total bacterial population. It means that, according to the equation of the linear regression, a P/HAc ratio value of 0.51 would be predicted if only PAO would be present. Either the observed highest anaerobic P/HAc ratio or the predicted value was similar to the stoichiometic ratio from a theoretical metabolic model for PAO cultures proposed by Smolders *et al.* (1994a) (0.50 P-mol/HAc C-mol). However, it does not match with higher P/HAc ratios (0.60 – 0.75 P-mol/HAc C-mol) reported elsewhere in the literature (Liu *et al.*, 1997; Sudiana *et al.*, 1999; Schuler and Jenkins, 2003).

Specific P-release rates and total P-released from each anaerobic batch test also showed strong correlations ($R^2 > 0.97$) with actual PAO and GAO fractions (Figures 6.5c and 6.5d). Therefore, considering the observed ratios in the present study, reliable relationships were made between observed biomass activity and actual PAO/GAO population ratios.

6.3.4. Anaerobic carbon transformations

The anaerobic conversions of relevant storage compounds were evaluated for each batch test. Figure 6.6 presents the stoichiometric ratios measured in the different batch tests and their trend line (black solid line). Using the theoretical anaerobic models proposed by Smolders *et al.* (1994a) and Zeng *et al.*, (2003a) for PAO and GAO, respectively, the theoretical anaerobic carbon transformations, according to actual PAO and GAO fractions present in each batch test, were calculated and also included in Figure 6.6 (dashed line).

Glycogen/HAc ratios observed in the 0/100, 20/80 and 40/60 PAO/GAO ratios batch tests are in the range of the values calculated with the theoretical models (Figure 6.6a). However, due to still unknown reasons, the rest of the ratios were slightly higher than the expected theoretical values. Gly/HAc ratios around 0.50 C-mol/C-mol have been reported for enriched PAO cultures in the literature (Smolders *et al.*, 1994b; Filipe *et al.*, 2001b; Schuler and Jenkins, 2003; Lu *et al.*, 2006). Considering the high enrichment of the PAO culture, it is unclear what caused a consistently slightly higher Gly/HAc ratio. However, the deviations are small compared to the observed standard deviations.

PHB/HAc ratios measured in this study, for the different PAO/GAO fractions, were in the range of the values calculated with the theoretical anaerobic models according to the PAO and GAO fractions studied in each batch test (Figure 6.6b).

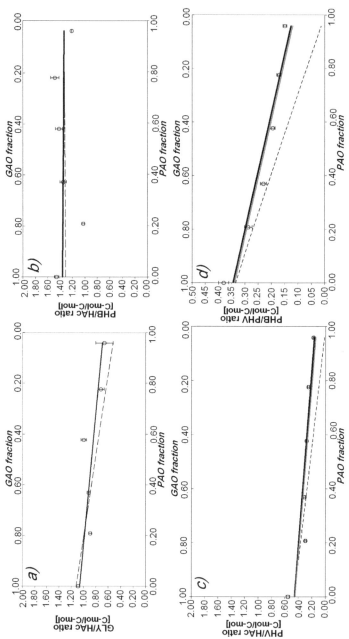

Figure 6.6. Anaerobic carbon transformations measured in the six different anaerobic PAO/GAO tests in terms of the PAO and GAO biomass fractions: a) Glycogen/HAc ratio, b) PHB/HAc ratio, c) PHV/HAc ratio and d) PHV/PHB ratio. Black solid line: experimental carbon transformations; dashed line: theoretical ratios and, gray solid line: theoretical ratios when a 10 % PHV production of total PHA in the theoretical PAO model was considered.

Figures 6.6c and 6.6d show the PHV/HAc and PHV/PHB ratios, respectively, observed in this study. There was a clear difference between experimental (black solid line) and theoretical values (dashed line). It was caused by the fact that PAO anaerobic metabolic model does not take into account PHV production despite that this storage compound was found in the current as well as in previous PAO cultures. Smolders *et al.* (1994a), Schuler and Jenkins (2003), Filipe *et al.* (2001b) and Lu *et al.* (2006) observed at least a small fraction of PHV (around 10 % of total PHA) in enriched PAO cultures. It was regularly assumed that a small GAO fraction present in PAO enriched cultures might lead to a low PHV production since, theoretically, PHV production is not involved in PAO's anaerobic metabolism. However, in this study, the GAO fraction present in the enriched PAO culture could be considered negligible (< 4 %). Therefore, based on previous experimental observations and assuming that PHV comprises 10 % of total PHA produced by PAO, a better fit between experimental (black solid line) and theoretical ratios (gray solid line) was obtained. Pereira *et al.* (1996) suggested a metabolic pathway by which PHV may be produced as part of the PAO metabolism. They claimed that it is the TCA cycle that still runs. Mino *et al.* (1998) argued that it is not correct due to the absence of a terminal electron acceptor and suggested that the glyoxylate pathway may be the reason. Kortstee *et al.* (2000) came with a different reasoning proposing the propionate-succinate pathway. PHV is, from our perspective, the result of metabolic side reactions (such as maintenance) as discussed by Pereira *et al.* (1996), Mino *et al.* (1998) and Kortstee *et al.* (2000).

6.3.5. Description of the method

Following the reliable relationships and strong correlations obtained between observed biomass activity from each batch test and actual PAO and GAO fractions ($R^2 > 0.97$) a method for the quantification of the PAO and GAO populations is proposed. Initially, the PAO fraction, as fraction of total PAO and GAO biomass, were calculated using an expression derived from the correlation between anaerobic P/HAc ratios and PAO fractions previously presented in Figure 6.5b:

$$f_{PAO} = \frac{\left[\dfrac{P}{HAc}\right]_{Exp}^{Ana}}{0.51} \qquad (6.1)$$

Where f_{PAO} is the PAO fraction in terms of total PAO and GAO populations (PAO/(PAO+GAO)) and, $\left[\dfrac{P}{HAc}\right]^{Ana}_{Exp}$ is the experimental anaerobic P/HAc ratio measured in the batch tests, in P-mol/C-mol HAc.

GAO fractions were calculated as the remaining fraction of the total PAO and GAO populations or with the following expression:

$$f_{GAO} = \frac{0.51 - \left[\dfrac{P}{HAc}\right]^{Ana}_{Exp}}{0.51} \tag{6.2}$$

Where f_{GAO} is the GAO fraction related to total PAO and GAO populations (GAO/(PAO+GAO)).

Furthermore, the active biomass fraction comprised by PAO and GAO in the activated sludge system ($f^{NET}_{(PAO+GAO)}$), could be determined based on the ratio between the maximum acetate uptake rate measured in the corresponding anaerobic batch test ($q^{MAX}_{HAc,TEST}$) and the average maximum acetate uptake rate (166 ± 6 mgHAc/gVSS/h) observed in the anaerobic batch tests executed with the enriched PAO and GAO cultures from this study (Figure 6.5a):

$$f^{NET}_{(PAO+GAO)} = \frac{q^{MAX}_{HAc,TEST}}{166} \tag{6.3}$$

The decision of using the average maximum acetate uptake rate for the estimation of the PAO and GAO populations was also strongly influenced and supported by the similar uptake rates reported in the literature for enriched cultures (Table 6.1).

After the net PAO plus GAO fraction were determined ($f^{NET}_{(PAO+GAO)}$), the net PAO and net GAO fractions (as % of VSS) could be computed with the following expressions:

$$f^{NET}_{PAO} = 100 \cdot f_{PAO} \cdot f^{NET}_{PAO+GAO} \tag{6.4}$$

$$f^{NET}_{GAO} = 100 \cdot f_{GAO} \cdot f^{NET}_{PAO+GAO} \tag{6.5}$$

6.3.6. Validation

The PAO and GAO fractions (as *Accumulibacter* and *Competibacter*, respectively) present in 8 different municipal wwtp were determined via FISH analyses and also using the method proposed in the present study (Table 6.2). Quantifications performed with the proposed method were expressed as fractions of MLVSS whereas quantifications carried out with FISH images were expressed as fractions of the biomass that gave a positive signal to the EUB mix probe. PAO and GAO fractions determined with the proposed method, in average, comprised around 10 and 2 % of total biomass (as MLVSS), respectively. On the other hand, quantifications of PAO (*Accumulibacter*) and GAO (*Competibacter*) fractions executed with FISH analysis resulted in average PAO and GAO fractions of 9 and 1 %, respectively.

Independently of the results provided by the proposed method and the direct FISH quantification, the PAO fractions measured in the municipal wastewater treatment plants are in the range observed in previous full-scale studies for *Accumulibacter* populations (7 – 20 %) (Saunders *et al.*, 2003; Tykesson *et al.*, 2006). However, lower *Competibacter* fractions (2 ± 1 %) were observed compared to previous full-scale reports (1 – 12 %, Saunders *et al.*, 2003; 10 – 31 % Wong *et al.*, 2005; 5 – 10 % Tykesson *et al.*, 2006).

A reasonable agreement was observed between the PAO and GAO fractions quantified with the proposed method and when FISH was applied (Figure 6.7) suggesting that the present method is potentially suitable for the estimation of PAO and GAO fractions in activated sludge systems.

Table 6.2. Comparison between PAO and GAO populations quantified with the proposed method and direct quantification with FISH analysis.

WWTP	Experimental Anaerobic P/HAc ratio mg/mg	HAc uptake rate mg/gVSS/h [1]	PAO and GAO fractions quantified with this method					PAO and GAO fractions quantified by FISH [7]	
			f_{PAO} [2]	f_{GAO} [3]	$f^{NET}_{(PAO+GAO)}$ [4]	f^{NET}_{PAO} [5]	f^{NET}_{GAO} [6]	PAO fraction	GAO fraction
1	0.46	19	0.90	0.10	0.11	10	1	16 ± 6	2 ± 1
2	0.36	22	0.71	0.29	0.13	9	4	11 ± 3	2 ± 1
3	0.43	13	0.84	0.16	0.08	7	1	6 ± 1	0 ± 0
4	0.45	19	0.88	0.12	0.11	10	1	7 ± 2	1 ± 1
5	0.46	9	0.90	0.10	0.05	5	1	6 ± 2	1 ± 1
6	0.38	17	0.75	0.25	0.10	8	3	6 ± 1	1 ± 0
7	0.33	14	0.65	0.35	0.08	5	3	10 ± 3	2 ± 1
8	0.45	19	0.88	0.12	0.11	10	1	10 ± 2	4 ± 2

[1] Maximum acetate uptake rate obtained from anaerobic batch tests, $q^{MAX}_{HAc,TEST}$.
[2] Calculated with expression 1, expressed as [PAO/(PAO+GAO)].
[3] Calculated with expression 2, expressed as [GAO/(PAO+GAO)].
[4] Fraction of active net PAO and GAO fractions present in the activated sludge, calculated with expression 6.3.
[5] Net PAO fraction per MLVSS present in activated sludge, calculated with expression 6.4, expressed as % of PAO per MLVSS.
[6] Net GAO fraction per MLVSS present in activated sludge, calculated with expression 6.5, expressed as % of GAO per MLVSS.
[7] Fractions directly quantified with FISH analysis, expressed as % of PAO or GAO fractions per total biomass (as EUB).

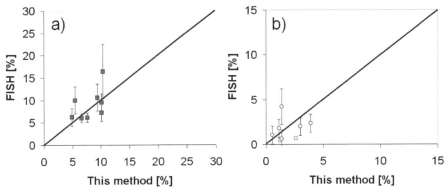

Figure 6.7. Comparison between (a) PAO and (b) GAO fractions from eight different full-scale EBPR activated sludge treatment plants quantified with the proposed method and by FISH.

6.4. Discussion

6.4.1. Anaerobic transformations in anaerobic PAO/GAO tests

Gly/HAc and PHB/HAc anaerobic carbon transformations were in the range of theoretical values calculated with the anaerobic metabolic models considering the actual PAO and GAO fractions present in each batch test (Figures 6.6a and 6.6b). However, PHV/HAc and PHV/PHB ratios showed an important difference compared to the prediction of the models (Figures 6.6c and 6.6d). Experimental data were well described only after a 10 % PHV production per HAc consumed was considered in the anaerobic model of PAO, what is in accordance with previous experimental observations of different authors (Smolders *et al.* 1994a; Pereira *et al.*, 1996; Brdjanovic *et al.* 1997; Filipe *et al.*, 2001b; Schuler and Jenkins, 2003; Lopez *et al.*, 2006), but not explicitly considered in the theoretical metabolic models of PAO. These facts underline the need to consider the role of the metabolic pathway that leads to PHV production in PAO's metabolic models in order to get a better description and prediction of the fate of internal storage compounds.

6.4.2. Method for the quantification of PAO and GAO

Important and strong correlations ($R^2 > 0.97$) were observed between biomass activity and PAO-GAO ratios in the different anaerobic batch tests (Figure 6.5). Additionally, the observed anaerobic carbon transformations under different actual PAO-GAO ratios (Figure 6.6) were in the range of previous observations drawn from enriched PAO and GAO cultures. These facts led and support the proposal of a practical method to quantify the PAO and GAO fractions in activated sludge systems.

The present method relies on the determination of the anaerobic P-release/HAc ratio to quantify the relative PAO and GAO fractions. In previous studies, Tykesson *et al.* (2006) and Oehmen *et al.* (2007) observed correlations between the anaerobic P-release/HAc ratio and PAO fractions (*Accumulibacter*) quantified via FISH. In particular, Oehmen *et al.* (2007) compiled data from several lab-scale studies and found a good correlation between the P-release/VFA ratio and the abundance of *Accumulibacter* quantified via FISH (Figure 6.8). If the experimental data regarding the correlation between P/HAc ratio and *Accumulibacter* fractions observed in this study are included (black squares and black solid line) among the data compiled by Oehmen *et al.* (2007) a good correlation covering all the experimental data is obtained (Figure 6.8, red solid line, $R^2 = 0.72$). It is remarkable that the slope of the correlation found in the present study (black solid line, 0.51) is comparable to the slope of the correlation that covers all the experimental data (red solid line, 0.53). This supports the use of the anaerobic P-release/HAc ratio for the quantification of the PAO and GAO fractions and consequently the validity of the present proposed method.

The proposed method for the quantification of PAO and GAO (as *Accumulibacter* and *Competibacter*, respectively) was evaluated with activated sludge from full-scale wastewater treatment plants. Quantifications performed with this laboratory research method compared to direct quantifications based on FISH analysis showed that the proposed method is suitable to estimate the PAO and GAO populations based on common standard determination methods (Figure 6.7).

The present method relies on previously proposed and experimentally validated anaerobic metabolic models for PAO and GAO. In particular, the PAO model has shown its reliability at full-scale systems (van Veldhuizen *et al.*, 1999; Brdjanovic *et al.*, 2000; Meijer *et al.*, 2002). On the other hand, despite that the GAO metabolic model (Zeng *et al.*, 2003a) has not been applied at full-scale systems it could be assumed that similar results at full-scale, when applied, could be expected since it followed a similar formulation as the PAO model.

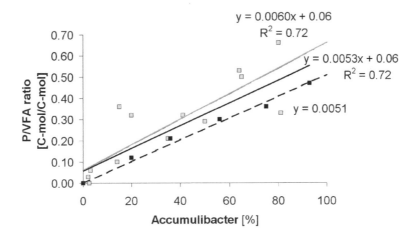

Figure 6.8. Correlations between experimental P-release/VFA ratios and abundance of *Accumulibacter* reported in the literature from different lab-scale studies performed under similar operating conditions (modified from Oehmen *et al.*, 2007). Gray squares and gray solid line: P-release/VFA ratios and correlation line as compiled and calculated by Oehmen *et al.* (2007); black squares and black solid line: experimental data and correlation observed in this study; and, red solid line: correlation between P-release/VFA ratios and *Accumulibacter* fractions after including the results from this study.

This proposed method, as all methods and models, has its own boundaries. Firstly, it is important to underline that, despite that it was developed using highly enriched cultures of PAO and GAO, these cultures were cultivated using a single carbon source (acetate). Acetate is the dominant VFA present in raw influents of full-scale municipal wastewater treatment plans. However, its presence and concentrations as well as of other types of VFA could vary from plant to plant affecting how PAO and GAO populations are exposed to different VFA fractions. Oehmen *et al.* (2005) found that, in short-term experiments, PAO were able to take up acetate practically at the same rate after switching from propionate to acetate while GAO required certain time before taking up acetate. It is noteworthy that despite all the different possible carbon sources present in full-scale systems an acceptable quantification of PAO (*Accumulibacter*) and GAO (*Competibacter*) was obtained with the proposed method (Figure 6.7).

The use of FISH for the quantification of PAO and GAO populations, not only from the enriched laboratory cultures but also from the full-scale activated sludge plants, are subjected to the limitations, inaccuracies, deviations and uncertainties derived from the technique. Mostly, probe specificity as well as presence of autofluorescent cells or debris may lead to either underestimation or overestimation of populations. It cannot be excluded that other PAO and GAO may be present in the systems that are not targeted by the present FISH analysis. Based on statistical estimations, the number of separate images needed in this study to quantify the PAO and GAO populations via FISH and obtain results within a 95 % confidence interval was 300. This number of images is required due to the high variability found among activated sludge flocs. To carry out an extensive quantification through FISH images was out of the scope of the present study due to time and financial limitations. Therefore, around 30 separate images were used to determine the corresponding PAO (as *Accumulibacter*) and GAO (as *Competibacter*) populations in the lab- and full-scale samples.

Maximum experimental anaerobic P/HAc ratio observed in this study was 0.47 P-mol/C-mol while a P/HAc ratio of 0.51 P-mol/C-mol was predicted if only PAO would be present. These values are in the range of the theoretical P/HAc ratios proposed and experimentally validated by Smolders *et al.* (1994a). However, they differed from higher ratios reported by other authors (0.60 – 0.75) (Liu *et al.*, 1997; Sudiana *et al.* 1999; Filipe *et al.*, 2001b; Schuler and Jenkins, 2003). A direct explanation to this difference cannot be given. It might partially be explained on the basis of the higher pH applied in those studies (> 7.2). Perhaps higher pH levels increased the anaerobic energy requirements for substrate uptake and maintenance leading to higher P/HAc ratios as found by Smolders *et al.* (1994a), Filipe *et al.* (2001b) and Schuler and Jenkins (2002). On the other hand, Liu *et al.* (1997) and Schuler and Jenkins (2003) reported that higher influent PO_4^{3-}-P/HAc ratios led to higher anaerobic P/HAc ratios. Therefore, Schuler and Jenkins (2003) suggested to using higher influent PO_4^{3-}-P/HAc ratios (> 0.12 P-mol/C-mol) in order to reach highly lab-enriched PAO cultures. An influent PO_4^{3-}-P/HAc ratio of 0.04 P-mol/C-mol was used in this study resulting in a highly enriched PAO culture (around 93 %). Similar influent ratios have been used in other studies reaching comparable enrichment levels (91 %, Lu *et al.*, 2006). This could indicate that a higher influent P/HAc ratio may favor PAO's enrichment but that it is not a strict requirement. Moreover, for the applicability of the present method, if higher anaerobic P/HAc ratios would have been observed in the different anaerobic lab-tests they might have led to an underestimation of PAO and overestimation of GAO populations (as *Accumulibacter* and *Competibacter*) while, using the found relationships, a good estimation of both microorganisms was obtained at full-scale systems (Table 6.2 and Figure 6.7).

According to Brdjanovic *et al.* (1997, 1998) and Lopez-Vazquez *et al.* (2007), PAO and GAO anaerobic stoichiometries are insensitive to temperature changes. Therefore, potentially PAO and GAO ratios could be reliably quantified using this method (at 20 $^{\circ}$C) regardless what the actual operating temperature is at the lab- or full-scale system.

6.4.3. Recommendations

A method has been proposed to quantify the PAO and GAO populations present in activated sludge systems. This method relies on the execution of a lab-batch experiment where HAc is added in excess and the activity is followed through HAc consumption and PO_4^{3-}-P release profiles. It is suggested to perform the anaerobic batch tests for a minimum period of 3 or 4 h to ensure that PAO and GAO are able to take up all possible HAc depending on their storage compound pools but without causing total HAc depletion. Preliminary batch tests executed with the activated sludge to be evaluated could be undertaken to define the minimum HAc concentration to perform the tests.

When this method is applied, it is suggested to keep a good pH and temperature control at 7.0 ± 0.05 and 20 ± 0.5 $^{\circ}$C, respectively. Preferably, pH and temperature fluctuations should be avoided to reduce the possibility of deviations. In addition, it must be warranted that, when anaerobic batch tests are executed, no other carbon sources rather than the acetate supplied at the beginning of the test is present. Besides this suggestion, nitrate or nitrite must be absent since they could favor the acetate uptake by ordinary denitrifying heterotrophs. In order to decrease the risk that other carbon sources are present or nitrate/nitrite intrusion, fresh mixed liquor (< 24 h older) should be used and biomass should be washed and resuspended 2 or 3 times in a synthetic mineral solution (further details can be found in Kuba *et al.*, 1996 and Wachtmeister *et al.* 1999). In addition biomass should be aerated for 1 – 1.5 h before test starts. Prior to aeration, ATU at a 20 mg/L concentration should be added in order to avoid the presence of nitrate or nitrite during the anaerobic batch tests due to nitrifying activity.

6.5. Conclusions

A simple method for the quantification of PAO and GAO population fractions in activated sludge systems has been proposed. It is based on reliable relationships observed in anaerobic batch tests between different PAO/GAO population ratios and biomass activity. For the determination of PAO and GAO fractions, the maximum anaerobic acetate uptake rate and anaerobic P-release/HAc ratio obtained from an anaerobic batch test executed with acetate in excess are required. The laboratory test was evaluated with activated sludge from municipal wastewater treatment plants, where the quantification performed following the proposed method was evaluated and compared to direct population estimations based on FISH measurements. The method showed to be potentially suitable to estimate the PAO and GAO populations regarding the total PAO-GAO biomass. It could be used not only to evaluate the performance of EBPR systems but also in the calibration of activated sludge models regarding the PAO-GAO interaction. It therefore is a good a practical protocol to be used to characterize EBPR wastewater systems in practice.

Acknowledgements

The authors would like to acknowledge the Mexican National Council for Science and Technology (CONACYT, Mexico) and the Autonomous University of the State of Mexico for the scholarship awarded to Carlos Manuel Lopez Vazquez. The valuable assistance provided by Mario Pronk, Gert van der Steen, Margarida Temudo and Udo van Dongen from Delft University of Technology for the analytical determination of storage compounds and FISH analysis is highly appreciated. Special thanks to the lab-staff from UNESCO-IHE Institute for Water Education.

References

APHA/AWWA/WPCF (1995) Standard methods for the examination of water and wastewater, 19th ed. Port City Press, Baltimore.

Amann RI (1995) In situ identification of microorganisms by whole cell hybridization with rRNA-targeted nucleic acid probes. In: Akkermans, A.D.L.; van Elsas, J.D.; de Bruijn, F.J. (Eds.), Molecular Microbial Ecology Manual. Kluwer Academic Publications, Dordrecht, The Netherlands.

Brdjanovic D, Logemann S, van Loosdrecht MCM, Hooijmans CM, Alaerts GJ, Heijnen JJ (1998) Influence of temperature on biological phosphorus removal: process and molecular ecological studies. Water Res., 32 (4), 1035.

Brdjanovic D, van Loosdrecht MCM, Versteeg P, Hooijmans CM, Alaerts GJ, Heijnen JJ (2000) Modeling COD, N and P removal in a full-scale wwtp Haarlem Waarderpolder. Water Res., 34 (3), 846.

Brdjanovic D, van Loosdrecht MCM, Hooijmans CM, Alaerts GJ, Heijnen JJ (1997) Temperature effects on physiology of biological phosphorus removal. A. S. C. E. J. Environ. Eng., 123 (2),144.

Brdjanovic D, van Loosdrecht MCM, Hooijmans CM, Mino T, Alaerts GJ, Heijnen JJ (1999) Innovative methods for sludge characterization in biological phosphorus removal systems. Water Sci. Technol., 39 (6), 37.

Cech JS, Hartman P (1993) Competition between phosphate and polysaccharide accumulating bacteria in enhanced biological phosphorus removal systems. Water Res., 27, 1219.

Crocetti GR, Banfield JF, Keller J, Bond PL, Blackall LL (2002) Glycogen accumulating organisms in laboratory-scale and full-scale wastewater treatment processes. Microbiol., 148, 3353.

Crocetti GR, Hugenholtz P, Bond PL, Schuler A, Keller J, Jenkins D, Blackall LL (2000) Identification of polyphosphate-accumulating organisms and design of 16S rRNA-directed probes for their detection and quantitation. Appl. Environ. Microbiol., 66 (3), 1175.

Filipe CDM, Daigger GT, Grady CPL Jr (2001a) A metabolic model for acetate uptake under anaerobic conditions by glycogen-accumulating organisms: stoichiometry, kinetics and effect of pH. Biotechnol. Bioeng., 76 (1), 17.

Filipe CDM, Daigger GT, Grady CPL Jr. (2001b) Stoichiometry and kinetics of acetate uptake under anaerobic conditions by an enriched culture of phosphorus-accumulating organisms at different pHs. Biotechnol. Bioeng., 76 (1), 32.

Henze M, Gujer W, Mino T, van Loosdrecht MCM (2000) Activated sludge models ASM1, ASM2, ASM2d and ASM3. I. W. A. Scientific and Technical Report No. 9. I. W. A. Task Group on Mathematical Modelling for Design and Operation of Biological Wastewater Treatment. I .W. A. Publishing: London.

Kortstee GJJ, Appeldoorn KJ, Bonting CFC, Van Niel EWJ, Van Veen HW (2000) Recent Developments in the Biochemistry and Ecology of Enhanced Biological Phosphorus Removal. Biochemistry (Moscow), 65 (3), 332.

Kuba T, M,urnleitner E, van Loosdrecht MCM, Heijnen JJ (1996) A metabolic model for the biological phosphorus removal by denitrifying organisms. Biotechnol. Bioeng., 52, 685.

Kuba T, Wachtmeister A, van Loosdrecht MCM, Heijnen JJ (1994) Effect of nitrate on phosphorus release in biological phosphorus removal systems. Water Sci Technol. 30, 263.

Lee N, Nielsen PH, Andreasen KH, Juretshko S, Nielsen JL, Schleifer KH, Wagner M (1999) Combination of fluorescence in situ hybridization and microautoradiography – a new tool for structure-function analyses in microbial ecology. App. Environ. Microbiol., 65 (3), 1289.

Liu WT, Mino T, Nakamura K, Matsuo T (1994) Role of glycogen in acetate uptake and polyhydroxyalkanoate synthesis in anaerobic-aerobic activated sludge with a minimized polyphosphate content. J. Ferment. Bioeng., 77 (5), 535.

Liu WT, Nakamura K, Matsuo T, Mino T (1997) Internal energy-based competition between polyphosphate- and glycogen-accumulating bacteria in biological phosphorus removal reactors-effect of P/C feeding ratio. Water Res., 31 (6), 1430.

Lopez C, Pons MN, Morgenroth E (2006) Endogenous process during long-term starvation in activated sludge performing enhanced biological phosphorus removal. Water Res., 40, 1519.

Lopez-Vazquez CM, Song Y-Il, Hooijmans CM, Brdjanovic D, Moussa MS, Gijzen H van Loosdrecht, M. C. M. (2007) Short-term temperature effects tests on the anaerobic metabolism of glycogen accumulating organisms. Biotechnol. Bioeng. (in press). 97(3):483-495.

Lu H, Oehmen A, Virdis B, Keller J, Yuan Z (2006) Obtaining highly enriched cultures of *Candidatus Accumulibacter Phosphatis* through alternating carbon sources. Water Res., 40 (20), 3838.

Meijer SCF, van Loosdrecht MCM, Heijnen JJ (2002) Modelling the start-up of a full-scale biological nitrogen and phosphorus removing WWTP's. Water Res., 36 (19), 4667.

Mino T, van Loosdrecht MCM, Heijnen JJ (1998) Microbiology and biochemistry of the enhanced biological phosphorus removal process. Water Res., 32 (11), 3193.

Murnleitner E, Kuba T, van Loosdrecht MCM, Heijnen JJ (1997) An integrated metabolic model for the aerobic and denitrifying biological phosphorus removal. Biotechnol. Bioeng., 54 (5), 434.

Oehmen A, Yuan Z, Blackall LL, Keller J (2005) Comparison of acetate and propionate uptake by polyphosphate-accumulating organisms and glycogen-accumulating organisms. Biotechnol. Bioeng., 91 (2), 162.

Oehmen A, Lemos PC, Carvalho G, Yuan Z, Keller J, Blackall LL, Reis MAM (2007) Advances in enhanced biological phosphorus removal: from micro to macro scale. Water Res., 41(11), 2271.

Otsu N (1979) A threshold selection method from gray-level histograms. I. E. E. E. Trans. Sys. Man. Cyber., 9 (1), 62.

Pereira H, Lemos PC, Reis MAM, Crespo JPSG, Carrondo MJT, Santos H (1996) Model for carbon metabolism in biological phosphorus removal processes based on in vivo C-NMR labelling experiments.Water Res., 30 (9), 2128.

Rieger L, Koch G, Kuhni M, Gujer W, Siegrist H (2001) The EAWAG bio-P module for Activated Sludge Model No. 3. Water Res., 35, 3887.

Satoh H, Mino T, Matsuo T (1994) Deterioration of enhanced biological phosphorus removal by the domination of microorganisms without polyphosphate accumulation. Water Sci. Technol., 30 (6), 203.

Saunders AM, Oehmen A, Blackall LL, Yuan Z, Keller J (2003) The effect of GAO (glycogen accumulating organisms) on anaerobic carbon requirements in full-scale Australian EBPR (enhanced biological phosphorus removal) plants. Water Sci. Technol., 47 (11), 37.

Schuler AJ, Jenkins D (2003) Enhanced biological phosphorus removal from wastewater by biomass with different phosphorus contents, Part 1: Experimental results and comparison with metabolic models. Water Environ. Res., 75 (6), 485.

Schuler AJ, Jenkins D (2002) Effects of pH on enhanced biological phosphorus removal metabolisms. Water Sci. Technol., 46 (4-5), 171.

Smolders GJF, van Der Meij J, van Loosdrecht MCM, Heijnen JJ (1994a) Model of the anaerobic metabolism of the biological phosphorus removal processes: stoichiometry and pH influence. Biotechnol. Bioeng., 43, 461.

Smolders GJF, van Der Meij J, van Loosdrecht MCM, Heijnen JJ (1994b) Stoichiometric model of the aerobic metabolism of the biological phosphorus removal process. Biotechnol. Bioeng, 44, 837.

Sudiana IM, Mino T, Satoh H, Nakamura K, Matsuo T (1999) Metabolism of enhanced biological phosphorus removal and non-enhanced biological phosphorus removal sludge with acetate and glucose as carbon source. Water Sci. Tech., 39 (6), 29.

Thomas M, Wright P, Blackall LL, Urbain V, Keller J (2003) Optimization of Noosa BNR plant to improve performance and reduce operating costs. Water Sci. Technol., 47 (12), 141.

Tykesson E, Blackall LL, Kong Y, Nielsen PH, Jansen JC (2006) Applicability of experience from laboratory reactors with biological phosphorus removal in full-scale plants. Wat Sci Technol., 54 (1), 267.

van Veldhuizen HM, van Loosdrecht MCM, Heijnen JJ (1999) Modelling biological phosphorus and nitrogen removal in a full scale activated sludge process. Water Res., 33 (16), 3459.

Wachtmeister A, Kuba T, van Loosdrecht MCM, Heijnen JJ (1997) A sludge characterization assay for aerobic and denitrifying phosphorus removing sludge. Water Res., 31 (3), 471.

Wong MT, Mino T, Seviour RJ, Onuki M, Liu WT (2005) *In situ* identification and characterization of the microbial community structure of full-scale enhanced biological phosphorous removal plants in Japan. Water Res., 39 (13), 2901.

Yagci N, Insel G, Artan N, Orhon D (2004) Modelling and calibration of phosphate- and glycogen-accumulating organism competition for acetate uptake in a sequencing batch reactor. Water Sci. Technol., 50 (6), 241.

Yagci N, Insel G Tasli R, Artan N, Randall CW, Orhon D (2006) A new interpretation of ASM2d for modeling of SBR performance for enhanced biological phosphorus removal under different P/HAc ratios. Biotechnol. Bioeng., 93 (2), 258.

Zeng RJ, van Loosdrecht MCM, Yuan Z, Keller J (2002) Proposed modifications to metabolic model for glycogen accumulating organisms under anaerobic conditions. Biotechnol. Bioeng. 80 (3), 277-279.

Zeng RJ, van Loosdrecht MCM, Yuan Z, Keller J (2003a) Metabolic model for glycogen-accumulating organisms in anaerobic/aerobic activated sludge sludge systems. Biotechnol. Bioeng., 81(1), 92.

Zeng RJ, Yuan Z, Keller J (2003b) Model-based analysis of anaerobic acetate uptake by a mixed culture of polyphosphate-accumulating and glycogen-accumulating organisms. Biotech. Bioeng., 83(3), 293.

7

Chapter

Modelling the PAO-GAO competition: the effects of carbon source, pH and temperature

Content

This chapter has been published as:
    Lopez-Vazquez CM, Oehmen A, Zhiguo Y, Hooijmans CM, Brdjanovic D, Gijzen HJ, van Loosdrecht MCM (2008) Modeling the PAO-GAO competition: effects of carbon source, pH and temperature. Water Res, 43(2):450-62.

Abstract

The influence of different carbon sources (acetate to propionate ratios), temperature and pH levels on the competition between polyphosphate- and glycogen-accumulating organisms (PAO and GAO, respectively) was evaluated using a metabolic model that incorporated the carbon source, temperature and pH-dependences of these microorganisms. The model satisfactorily described the bacterial activity of PAO (*Accumulibacter*) and GAO (*Competibacter* and *Alphaproteobacteria-GAO*) laboratory-enriched cultures cultivated on propionate (HPr) and acetate (HAc) at standard conditions (20 °C and pH 7.0). Using the calibrated model, the effects of different influent HAc to HPr ratios (100-0, 75-25, 50-50 and 0-100 %), temperatures (10, 20 and 30 °C) and pH levels (6.0, 7.0 and 7.5) on the competition among *Accumulibacter, Competibacter* and *Alphaproteobacteria-GAO* were evaluated. The main aim was to assess which conditions were favorable for the existence of PAO and, therefore, beneficial for the biological phosphorus removal process in sewage treatment plants. At low temperature (10 °C), PAO were the dominant microorganisms regardless of the used influent carbon source or pH. At moderate temperature (20 °C), PAO dominated the competition when HAc and HPr were simultaneously supplied (75 - 25 and 50 - 50 % HAc to HPr ratios). However, the use of either HAc or HPr as sole carbon source at 20 °C was not favorable for PAO unless a high pH was used (7.5). Meanwhile, at higher temperature (30 °C), GAO tended to be the dominant microorganisms. Nevertheless, the combined presence of acetate and propionate in the influent (75 – 25 and 50 - 50 % HAc to HPr ratios) as well as a high pH (7.5) appear to be potential factors to favor the metabolism of PAO over GAO at higher sewage temperature (30 °C).

7.1. Introduction

In the pursuit for control strategies to suppress the proliferation of GAO, certain environmental and operational conditions have been identified as key factors to understand the competition between PAO and GAO: type of VFA present in the influent (e.g. acetate or propionate) (Pijuan *et al.*, 2004; Oehmen *et al.*, 2005a, 2006a), pH (Filipe *et al.*, 2001c; Oehmen *et al.*, 2005c), temperature (Whang and Park, 2006; Lopez-Vazquez *et al.*, 2007b; 2008b) and the phosphorus (P) to VFA ratio (Liu *et al.*, 1997). Nevertheless, little is known about the effects that the combination of these factors may have on the PAO-GAO competition.

Nowadays, mathematical modeling has become an integral part of biological wastewater treatment, often for optimization and prediction of process performance, and as a supporting tool for design. Recently, modeling techniques have also been applied to obtain a more accurate evaluation and understanding of the complex microbial interactions that take place in activated sludge systems. Within this context, Manga *et al.* (2001), Zeng *et al.* (2003b), Yagci *et al.* (2003, 2004) and Whang *et al.* (2007) have developed mathematical models aiming at improving the understanding concerning the interaction between PAO and GAO. However, these models have been built up based on a relatively narrow range of operational and/or environmental conditions, and the impact of one parameter on the PAO-GAO competition is evaluated independently of other factors. This limits their applicability, and does not allow the study and evaluation of the combined effects of key factors that influence the PAO-GAO competition, such as the carbon source, pH and temperature. On the other hand, to study the combined effects of different conditions through lab-scale experiments would require a considerable investment of time and resources, which may be assessed more rapidly and economically by mathematical modeling (Salem *et al.*, 2002). Thus, an integrated mechanistic mathematical model that takes into account the effect, dependence and influence of these factors on the metabolisms of PAO and GAO is clearly needed. This may be an important, flexible and useful tool towards the optimization of the EBPR process through improving the understanding of the interaction among bacterial populations under different environmental and operational conditions.

The main objective of the present study is to evaluate the combined effects of key operating conditions on the PAO-GAO competition using an integrated metabolic model. The different proposed metabolic models of PAO and GAO on acetate (HAc) and propionate (HPr) as well as the temperature and pH dependencies of these microorganisms obtained through several lab-scale studies were incorporated. Using the integrated metabolic model, certain scenarios were evaluated by combining different HAc to HPr ratios (100-0, 75-25, 50-50 and 0-100 %), pH levels (6.0, 7.0 and 7.5) and temperatures (10, 20 and 30 °C).

7.2. Materials and methods

7.2.1. Model description

The model was developed by incorporating, in an overall model, the metabolic models for PAO (as *Accumulibacter*) and GAO (as *Competibacter* and *Defluviicoccus*) on acetate (HAc) and propionate (HPr) (Smolders *et al.*, 1995; Filipe *et al.*, 2001a; Zeng *et al.*, 2003a; Oehmen *et al.*, 2005b, 2006b, 2007b) as well as their temperature (Brdjanovic *et al.*, 1997, 1998; Lopez-Vazquez *et al.*, 2007b, 2008b,c) and pH dependencies (Filipe *et al.*, 2001b,d) of their anaerobic and aerobic metabolisms. The integrated mathematical model describing the activity of PAO and GAO in a laboratory-scale sequencing batch reactor (SBR) was implemented in Aquasim (Reichert, 1994).

The model accounts for biomass distribution among 10 particulate components (Appendix 7.1). Three active biomass types include (Oehmen *et al.*, 2007a): *Accumulibacter* X_{PAO} (known PAO), *Competibacter* $X_{Competi}$ (known GAO) and GAO from the *Alphaproteobacteria* group X_{Alpha} (hereafter referred to as *Alphaproteobacteria-GAO*). The intracellular compounds stored by these three active biomass types were defined as poly-β-hydroxyalkanoates (PHA) (X_{PHA}^{PAO}, $X_{PHA}^{Competi}$ and X_{PHA}^{Alpha}); and, glycogen (X_{GLY}^{PAO}, $X_{GLY}^{Competi}$ and X_{GLY}^{Alpha}), while poly-phosphate (poly-P) (X_{PP}) was also defined for X_{PAO}. Further, the model describes the concentrations of the different PHA fractions stored in the anaerobic stages by the biomass: X_{PHB}^{PAO}, $X_{PHB}^{Competi}$, $X_{PHV}^{Competi}$, $X_{PH2MV}^{Competi}$, X_{PHB}^{Alpha}, X_{PHV}^{Alpha} and X_{PH2MV}^{Alpha} (Appendix 7.2). These anaerobically stored PHA fractions are essential components of the model for the determination of the biomass aerobic stoichiometry (Appendix 7.3); however, for simplification purposes, they are not explicitly displayed in the main stoichiometric matrix (Appendix 7.1). Four dissolved components are considered relevant for the biological conversions and process stoichiometry: oxygen (S_{O2}), acetate (S_{HAc}), propionate (S_{HPr}) and orthophosphate (PO4-P) (S_{PO4}). This led to a mathematical model with 19 processes and 21 components.

The aerobic stoichiometry and kinetics (Appendices 7.1 and 7.6) were described following the approach proposed by Murnleiter *et al.* (1997), where the biomass growth is result of the difference between the PHA oxidized minus the PHA utilized for poly-P and glycogen formation purposes, favoring the replenishment and formation of intracellular stored compounds over growth. Ecologically, the organisms have to survive in dynamic systems based on adequate levels of internally stored substrate and, to compete efficiently, they have to re-supply their storage pools quickly (Murnleitner *et al.*, 1997).

In order to decrease the number of variables influencing the PAO-GAO competition and provide a common platform for the evaluation of the combined effects of the key factors under study, the same δ value (or P/O ratio), and anaerobic (m_{ATP}^{ANA}) and aerobic maintenance coefficients (m_{ATP}^{AER}) (1.85 ATP-mol/NADH-mol, 2.35x10^{-3} ATP-mol/(C-mol-h) and 0.019 ATP-mol/(C-mol-h), respectively) were assumed for all microorganisms (Smolders *et al.* 1994b). Within the same context, also identical half-saturation coefficients for acetate and propionate ($K_{S,HAc}$ and $K_{S,HPr}$ of 0.001

mmol/mmol), intracellular storage products ($K_{S,PHA}$, $K_{S,GLY}$, and $K_{S,PP}$ of 0.01 mmol/mmol) and soluble compounds ($K_{S,PO4}$ and $K_{S,O2}$ of 0.01 mmol/mmol) were assigned to all microorganisms (Appendix 7.7).

7.2.1.1. Carbon source effects

The type of carbon source in the influent has direct effects on the anaerobic metabolisms of PAO and GAO and causes indirect effects on their aerobic metabolisms.

According to Oehmen *et al.* (2005a, 2006a,b), the type of carbon source, either HAc or HPr, has a direct effect on the anaerobic metabolism of PAO and GAO. *Accumulibacter* (a known PAO, Oehmen *et al.*, 2007a) are able to take up these two VFA with the same efficiency and at a similar kinetic rate (Oehmen *et al.*, 2005a, 2006a,b). On the other hand, at standard conditions (20 °C and 7.0 pH), *Competibacter* can store HAc at a similar rate as *Accumulibacter* but take up HPr at negligible rates, whereas the *Alphaproteobacteria-GAO* can take up HPr at a similar rate as *Accumulibacter* but take up HAc at about 50 % of their rate (Oehmen *et al.*, 2005a, 2006a,b). Furthermore, the uptake of either HAc or HPr leads to different total anaerobic PHA productions per C-mol of carbon source, and also to the storage of different PHA fractions: poly-β-hydroxybutyrate (PHB), poly-β-hydroxyvalerate (PHV) or poly-β-hydroxy-2-methylvalerate (PH₂MV). Thus, C-source effects on the anaerobic metabolism of PAO and GAO are taken into account by incorporating their proposed anaerobic stoichiometries on HAc and HPr as well as their observed maximum substrate uptake rates in the integrated metabolic model (Table 7.1, and Appendices 7.2 and 7.5).

The type of carbon source also has an indirect effect on the aerobic metabolism of PAO and GAO because their maximum aerobic yields depend upon the PHA fractions stored under anaerobic conditions. This can be explained in terms of the two molecules that have gone through the two initial steps of PHA production (e.g. condensation and reduction) also known as reduced PHA precursors: acetyl-CoA* and propionyl-CoA* (see Filipe *et al.*, 2001a; Zeng *et al.*, 2003a). This nomenclature is followed in order to keep the initial model development as simple as possible and avoid nonlinear processes. Thus, by making the maximum aerobic yields of PAO and GAO dependent upon δ, and the acetyl-CoA* and propionyl-CoA* fractions (Zeng *et al.*, 2003a), the aerobic metabolism of PAO and GAO can be modelled as function of the carbon sources taken up under anaerobic conditions (Appendixes 7.1, 7.3 and 7.4).

Table 7.1. Anaerobic stoichiometry and kinetics for *Accumulibacter, Competibacter* and *Alphaproteobacteria-GAO* as a function of the carbon source at standard conditions (pH 7.0 and 20 °C).

Microorganism	Carbon source	$q_{X,VFA}^{MAX}$ [a]	PHA production [C-mmol/C-mmol VFA]			
			PHB	PHV	PH2MV	Total PHA
Accumulibacter	Acetate[b]	0.20	1.33	0	0	1.33
	Propionate[c]	0.20	0	0.56	0.67	1.23
Competibacter	Acetate[d]	0.20	1.36	0.46	0.04	1.86
	Propionate[e]	0.01	0.10	0.58	0.82	1.50
Alphaproteobacteria-GAO	Acetate[e]	0.10	1.36	0.46	0.04	1.86
	Propionate [f]	0.20	0.10	0.58	0.82	1.50

[a]. Maximum substrate uptake rates were rounded to 0.20 and 0.10 C-mmol/(C-mmol/h) based on Lopez-Vazquez *et al.* (2007b) and this study.
[b]. Smolders *et al.* (1994a)
[c]. Oehmen *et al.* (2005b)
[d]. Zeng *et al.* (2003a)
[e]. Oehmen *et al.* (2005a)
[f]. Oehmen *et al.* (2006b)

7.2.1.2. Temperature effects

In order to incorporate the temperature effects into the currently proposed model, and thus extend its range and applicability, an assumption was made: that PAO (*Accumulibacter*) cultivated on HAc and HPr have the same temperature dependencies, and that both GAO grown on HAc and HPr (*Competibacter* and *Alphaproteobacteria-GAO*) share similar temperature dependencies as well. This assumption was made considering the successful application of mathematical models at full-scale activated sludge systems, regardless of the potential presence of different PAO strains (van Veldhuizen *et al.*, 1999; Brdjanovic *et al.*, 2000; Meijer *et al.*, 2002). This suggests that if there are different strains present, they appear to have similar temperature dependencies.

The temperature dependences of the anaerobic and aerobic metabolic processes determined by Brdjanovic *et al.* (1997, 1998) and Meijer *et al.* (2002) for PAO, and Lopez-Vazquez *et al.* (2007b, 2008b) for GAO were incorporated in order to describe the metabolic processes of these microorganisms from 10 to 30 °C (Table 7.2). The complete kinetic rate expressions for the 19 different metabolic processes of the model are presented in Appendix 7.5.

Table 7.2. Temperature coefficients for the anaerobic and aerobic metabolic processes of PAO and GAO.

Process	Organism	
	PAO (*Accumulibacter*)	**GAO** (*Competibacter* and *Alphaproteobacteria-GAO*)
Anaerobic acetate uptake rate	1.054 [a]	1.054 [b]
Anaerobic maintenance	1.096 [c]	1.028 [d]
PHA degradation	1.129 [e]	1.141 [f]
Glycogen production	1.125 [g]	1.090 [h]
Poly-P formation	1.031 [e]	-
Aerobic maintenance	1.064 [c]	1.054 [d]

[a]. Lopez-Vazquez *et al.* (2007b), valid from 10 to 20 °C.
[b]. Lopez-Vazquez *et al.* (2007b). Valid from 10 to 40 °C in combination with the inactivation expression.
[c]. Brdjanovic *et al.* (1997), valid from 5 to 30 °C.
[d]. Lopez-Vazquez *et al.* (2007b), valid from 10 to 40 °C.
[e]. Adjusted in this study based on the observations of Brdjanovic *et al.* (1998), valid from 5 to 20 °C.
[f]. Lopez-Vazquez *et al.* (2008c), valid from 10 to 30 °C.
[g]. Meijer *et al.* (2002), valid from 10 to 30 °C.
[h]. Lopez-Vazquez *et al.* (2008b), valid from 10 to 40 °C in combination with its inactivation expression.

7.2.1.3. pH effects

pH affects the anaerobic and aerobic metabolisms of PAO and GAO in different fashions. The effect of pH on the anaerobic metabolism of PAO and GAO was accounted for through: (a) the α parameter, which represents the energy (ATP) necessary for the transport of substrate over the cell membrane, and is assumed to be generated through the hydrolysis of poly-P and glycogen for PAO and GAO, respectively (Smolders *et al.*, 1994a; Filipe *et al.*, 2001a,d), (see Appendix 7.2); (b) the non-linear pH-dependency of the anaerobic stoichiometry of PAO with HPr (Appendix 7.2), as described by Oehmen *et al.* (2005c); and (c) the pH-effect on the anaerobic substrate uptake rates of GAO. A substantial decrease in the substrate uptake rates of GAO as pH rises from 6.0 to 7.5 was observed by Filipe *et al.* (2001a), whereas the substrate uptake rate of PAO was insensitive to pH changes

(Filipe *et al.*, 2001d). The substrate uptake rate dependency of GAO on pH was taken into account by incorporating a Monod-type expression developed by Filipe *et al.* (2001a) but adjusted to obtain a 0.20 C-mmol/(C-mmol-h) substrate uptake rate at pH 7.0 and 20 °C (Appendix 7.5). Figure 7.1 shows the combined effects of pH and temperature on the anaerobic acetate uptake rates of PAO and GAO (Figure 7.1a) as well as the combined effects on their anaerobic propionate uptake rates (Figure 7.1b). As discussed above for the case of temperature, it was also assumed that GAO grown on HAc and HPr share similar pH dependencies.

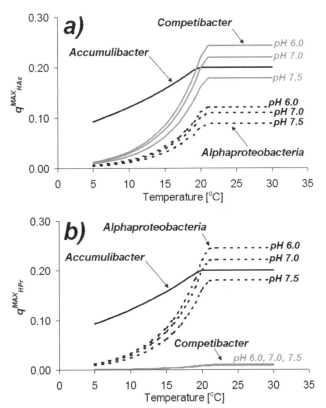

Figure 7.1. Combined temperature and pH effects on the maximum substrate uptake rates of *Accumulibacter, Competibacter* and *Alphaproteobacteria-GAO* for (a) acetate and (b) propionate. Black solid line: *Accumulibacter*; gray solid lines: *Competibacter*; and, black dotted lines: *Alphaproteobacteria-GAO*. Maximum substrate uptake rates of *Accumulibacter* are insensitive to pH changes.

The aerobic kinetic parameters of PAO and GAO incorporated in the current model are independent of the pH since, according to Filipe *et al.* (2001b), the aerobic metabolism of PAO and GAO are insensitive to pH changes between 6.0 and 7.5.

7.2.2. Modeling the operation of the lab-scale reactors

In order to model and simulate the hydraulic operation of the lab-scale sequencing batch reactor (SBR), as well as its alternating anaerobic-aerobic-settling-idle stages, a similar approach as Moussa *et al.* (2006) was adopted, by defining the reactor in Aquasim (Reichert *et al.*, 1994) as a mixed reactor compartment with variable volume. Thereby, the integrated metabolic model was incorporated into the model of the SBR.

7.2.3. Model calibration and validation

The integrated metabolic model was calibrated by fitting the PHA, glycogen and PO_4-P profiles of the enriched PAO and GAO cultures cultivated by Oehmen *et al.* (2005a,b; 2006b; 2007b) and Lopez-Vazquez *et al.* (2007b) at standard conditions (pH 7.0 and 20 $^\circ$C). Only the aerobic PHA degradation (k_{PHA}), glycogen formation (k_{GLY}) and poly-P formation rates (k_{PP}) were the kinetic parameters adjusted to calibrate the model. In order to describe the steady-state bacterial activities observed in the PAO and GAO reactors, simulations for an equivalent SBR operation period of 25 or 30 days were executed. This simulation period corresponded to approximately 3 times the solids retention time (SRT) applied for the enrichment of the cultures (7 or 10 days, depending on the reactor).

The calibrated model was validated by simulating the steady-state bacterial activities observed in different cycle measurements from the enriched cultures of Oehmen *et al.* (2004) and Lopez-Vazquez *et al.* (2007a). Simulations were executed for an equivalent operational period of at least 25 days.

7.2.4. Simulating the combined effect of environmental and operational conditions

The combined effects of different HAc to HPr ratios (100-0, 75-25, 50-50 and 0-100 %), pH levels (6.0, 7.0 and 7.5) and temperatures (10, 20 and 30 $^\circ$C) on the PAO-GAO competition were evaluated using the calibrated metabolic model. This led to the execution of 36 different simulations. The simulated synthetic influent contained: 5 C-mmol/L of VFA as HAc and/or HPr depending on the studied HAc to HPr ratio, and 0.40 P-mmol of orthophosphate (approximately 25 mg/L PO_4^{3-}-P). This resulted in a P/VFA ratio of 0.08 P-mmol/C-mmol. At the beginning of each simulation, the initial active biomass concentrations (X_{PAO}, $X_{Competi}$ and X_{Alpha}) were identical (30 C-mmol/L). In order to avoid limiting conditions at the start of the simulations, the initial storage fractions (f_{GLY} and f_{PP}, Appendix 7.7) were similar to the maximum fractions in biomass (f_{GLY}^{MAX} and f_{PP}^{MAX}) (Appendix 7.7). Mathematical simulations were executed for an equivalent SBR operational period of 50 days, which is slightly longer than six times the applied SRT of 8 days.

7.3. Results

7.3.1. Calibrated model

In order to describe the anaerobic conversions, the anaerobic maximum substrate uptake rates reported in previous studies were incorporated into the developed model (Oehmen *et al.*, 2005a,b, 2006b, 2007b; Lopez-Vazquez *et al.*, 2007b). For the aerobic conversions, the integrated model was calibrated by adjusting the aerobic coefficients of the studied organisms (k_{PHA} and k_{GLY}, and, for PAO, also k_{PP}). The calibrated model was able to satisfactorily describe the biomass activities observed in the enriched PAO and GAO cultures (Oehmen *et al.*, 2005a,b, 2006b, 2007b; Lopez-Vazquez *et al.*, 2007b) (Figure 7.2).

The active biomass concentrations computed by the calibrated model, after the mathematical simulations reached steady-state conditions, were similar to the biomass concentrations experimentally quantified in the different lab-scale SBR (Oehmen *et al.*, 2005a,b, 2006b, 2007b; Lopez-Vazquez *et al.*, 2007b) (Table 7.3). Therefore, the calibrated model was not only able to describe the activity of the studied organisms under different carbon sources but also able to provide a good prediction of the biomass concentrations observed in the corresponding cycle measurements, thus confirming its reliability.

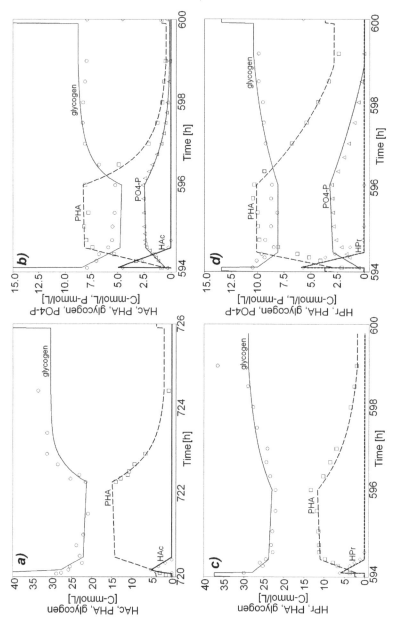

Figure 7.2. Calibrated model describing the biological activity observed in (a) an enriched GAO culture on acetate (*Competibacter*); (b) an enriched PAO culture on acetate (*Accumulibacter*); (c) an enriched GAO culture on propionate (*Alphaproteobacteria-GAO*); and (d) an enriched PAO culture on propionate (*Accumulibacter*). Measured parameters: carbon source, acetate or propionate (◊); PHA (□); glycogen (○); and, orthophosphate (PO4-P) (Δ). Lines indicate model description; bold line: acetate; short dashed line: propionate; dashed line: PHA; solid line: glycogen; dotted line: orthophosphosphate (PO4-P).

Table 7.3. Comparison between measured and predicted biomass concentrations.

Microorganism	Carbon source	Biomass concentration		
		Predicted [C-mmol/L]	Measured [C-mmol/L]	Difference [%]
Accumulibacter	Acetate	54	51	6
Accumulibacter	Propionate	75	67	12
Competibacter	Acetate	87	96 ± 4	9
Alphaproteobacteria-GAO	Propionate	91	84 ± 3	8

The calibrated aerobic kinetic parameters for the three different bacterial populations are displayed in Table 7.4. The calibrated aerobic kinetic rates of PAO on HAc differed from those of PAO on HPr. Consequently, the aerobic kinetic rates of PAO were expressed as a function of the HAc to HPr ratio (Appendix 7.5), using the different aerobic kinetic parameters determined for the cases of HAc and HPr as the sole carbon source (Table 7.4). This approach enables the modeling of the PAO activity under the simultaneous presence of these two VFA.

Table 7.4. Comparison of the adjusted aerobic kinetic parameters for the model calibration.

Aerobic kinetic parameter [a]		Organism					
		Accumulibacter		*Competibacter*		*Alphaproteobacteria GAO*	
		HAc	HPr	HAc	HPr	HAc	HPr
PHA degradation	k_{PHA}	0.800	0.330	1.100	-	1.100	0.500
Glycogen production	k_{GLY}	0.015	0.008	0.300	-	0.300	0.060
Poly-P formation	k_{PP}	0.020	0.002	-	-	-	-

[a] C-mol/(C-mol-h) units.

Since *Competibacter* and *Alphaproteobacteria-GAO* cultivated on HAc show similar anaerobic stoichiometry and kinetics (Dai *et al.*, 2007), the calibrated aerobic kinetic parameters of *Competibacter* on HAc were also used to describe the aerobic activity of *Alphaproteobacteria-GAO* on HAc (Table 7.4).

7.3.2. Model validation

The model was validated using different data sets from enriched PAO and GAO cultures (Oehmen *et al.*, 2006; Lopez-Vazquez *et al.*, 2007a) than the ones used for the calibration. A good description of the activity of the bacterial communities was observed (Figure 7.3). However, despite that most of the parameters were well-described, glycogen was often overestimated. The differences between the observed and modeled glycogen concentrations were up to 20-25 % in the case of the enriched PAO cultures (Figure 7.3). In previous models, the description of glycogen was also observed to be one of the most sensitive processes (van Veldhuizen *et al.*, 1999; Brdjanovic *et al.*, 2000; Meijer *et al.*, 2002).

7.3.3. Modeling the carbon, temperature and pH effects

Thirty-six simulations were executed using the calibrated metabolic model, aiming at studying the influence of different HAc to HPr ratios (100-0, 75-25, 50-50 and 0-100 %), pH levels (6.0, 7.0 and 7.5) and temperatures (10, 20 and 30 °C) on the PAO-GAO competition. Figure 7.4 shows the predicted biomass fractions from the different executed simulations (as percentage of the total biomass concentration) after an equivalent SBR operating time of 50 days (which corresponds to approximately 6 times the SRT). After this simulation period, the main trend of the biomass fractions hardly changed (data not shown) indicating that the simulation time was sufficient to define the dominance of any of the microorganisms or their coexistence.

When HAc was used as the sole carbon source, *Accumulibacter* dominated the system (biomass fractions higher than 60 %) at low temperature (10 °C) (Figures 4a, 4b and 4c). *Accumulibacter* were also the dominant microorganisms at 20 °C when a high pH was applied (7.5) (Figure 7.4c). At 20 °C and lower pH levels (pH 6.0 and 7.0), *Accumulibacter* tended to coexist with *Competibacter* (Figures 4a and 4b). At higher temperature (30 °C), *Competibacter* were predominant independently of the applied pH (Figures 4a, 4b and 4c). Similar biomass distributions were observed between *Accumulibacter* and *Alphaproteobacteria-GAO* when propionate was the sole carbon source (100 % HPr), (Figures 4j, 4k and 4l).

When HAc and HPr were simultaneously supplied (75-25 and 50-50 % HAc to HPr ratios), *Accumulibacter* were the dominant microorganisms at low and moderate temperature (10 and 20 °C), regardless of the applied pH (Figures 4d – 4i). At 30 °C and with an influent 75 - 25 % HAc to HPr ratio, *Accumulibacter* tended to dominate the system only if a 7.0 or 7.5 pH was applied (Figures 4e and 4f). Whereas, when a 50 – 50 % HAc to HPr ratio was supplied at 30 °C, a high pH (7.5) had to be used to favor *Accumulibacter* over GAO (Figure 7.4i).

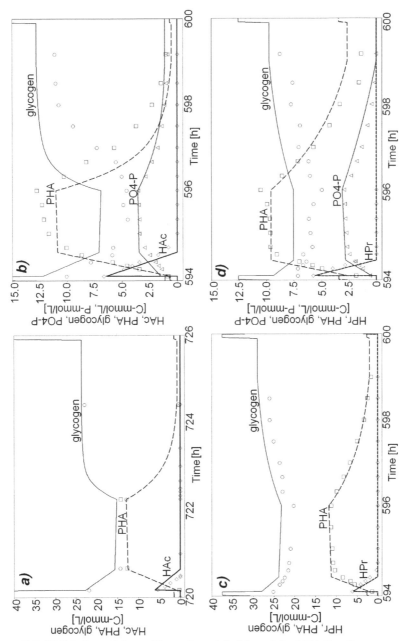

Figure 7.3. Validation of the calibrated model: (a) enriched GAO culture on acetate (*Competibacter*); (b) enriched PAO culture on acetate (*Accumulibacter*); (c) enriched GAO culture on propionate (*Alphaproteobacteria-GAO*); and (d) enriched PAO culture on propionate (*Accumulibacter*). Measured parameters: carbon source, acetate or propionate (◊); PHA (□); glycogen (○); and, orthophosphate (PO4-P) (Δ). Lines indicate model description; bold line: acetate; short dashed line: propionate; dashed line: PHA; solid line: glycogen; dotted line: orthophosphosphate (PO4-P).

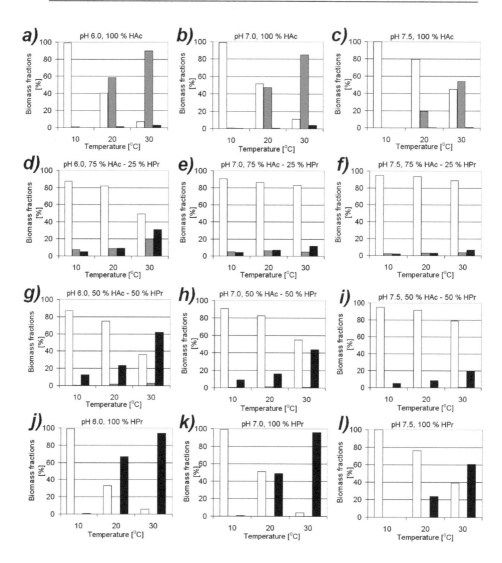

Figure 7.4. Bacterial population distributions as a function of the carbon source (acetate to propionate ratios), pH and temperature. *Accumulibacter* (PAO): white; *Competibacter* (GAO): dark grey; and, *Alphaproteobacteria* (GAO): black.

7.4. Discussion

7.4.1. Effect of the δ-ratio

An identical δ-ratio of 1.85 (Smolders *et al.*, 1994b) was assigned to all organisms in order to model their aerobic metabolisms. This ratio is a measure of the efficiency of the oxidative phosphorylation, indicating the energy produced (as ATP) per nicotinamide adenine dinucleotide ($NADH_2$) oxidized. All aerobic yield coefficients depend on this metabolic ratio (Appendix 7.3). It has a maximal theoretical value of 3. For normal microbial processes, it usually ranges between 1.5 and 2.0 (Figure 7.5). A correlation between δ and m_{ATP}^{AER} for different types of mixed cultures enriched under diverse electron acceptors and donors (Figure 7.5) suggests that they cannot easily be estimated independently, as observed by Dircks *et al.* (2001). In this study, the δ-ratio of 1.85 ATP-mol/($NADH_2$-mol) determined by Smolders *et al.* (1994b) was used because this is the only ratio that has been directly calculated using an enriched biomass. Other δ-ratios reported in literature have been indirectly determined or estimated (Figure 7.5). In fact, only for enriched PAO cultures can the δ-ratio be directly determined since the aerobic metabolic processes can be decoupled from the aerobic maintenance requirements (Smolders *et al.*, 1994b). Unless δ can be directly estimated, we suggest using this value when modeling the aerobic metabolisms of mixed and enriched cultures in activated sludge systems. Nevertheless, it can not be discarded that, depending on their physiology or as consequence of the influence (and combination) of different environmental and operational factors on their cultivation, the δ value of certain microorganisms may be different (Dias *et al.*, 2008).

Simulations were carried out aiming at evaluating the influence of diverse δ-ratios on the competition between *Accumulibacter* and *Competibacter* on HAc, and *Accumulibacter* and *Alphaproteobacteria* on HPr at 20 °C and pH 7.0 (Table 7.5). The simulations were performed for 25 d (around 3 times the SRT) starting with identical initial biomass fractions. The use of different δ-ratios only had a marginal effect on the PAO and GAO fractions predicted by the model (Table 7.5). This implies that the use of an identical δ-ratio (1.85 ATP-mol/($NADH_2$-mol)) does not have a major effect when modeling the PAO-GAO competition. Moreover, using an identical δ makes the aerobic stoichiometries of the different microorganisms similar and, consequently, reduces the number of variables to evaluate. This makes the PAO-GAO competition only dependent upon the carbon sources, temperature and pH effects.

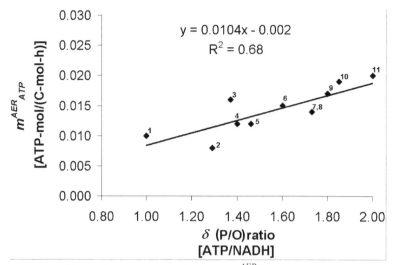

Figure 7.5. Aerobic maintenance coefficient (m_{ATP}^{AER}) as a function of the δ (P/O) ratio reported in literature. 1: Kuba et al. (1996); 2: Oehmen et al. (2006b); 3: Oehmen et al. (2007b); 4: Dircks et al. (2001); 5: Brdjanovic et al. (1997); 6: Beun et al. (2000a); 7: Zeng et al. (2003a); 8: Lopez-Vazquez et al. (2008b); 9: Dircks et al. (2001); 10: Smolders et al. (1994b); and, 11: Beun et al. (2000b).

Table 7.5. Net biomass yields of PAO and GAO as function of the carbon source and δ-ratios reported in the literature at standard conditions (20 °C, pH 7.0).

Carbon source	Organism	δ ratio	Net biomass growth [e]	Biomass [f] fractions [%]
Acetate (HAc)	*Accumulibacter*	1.85 [a]	0.33	52
	Competibacter	1.73 [b]	0.32	48
Propionate (HPr)	*Accumulibacter*	1.37 [c]	0.40	51
	Alphaproteobacteria-GAO	1.29 [d]	0.40	49
Acetate (HAc)	*Accumulibacter*	1.85 [g]	0.33	54
	Competibacter	1.85 [g]	0.33	46
Propionate (HPr)	*Accumulibacter*	1.85 [g]	0.48	54
	Alphaproteobacteria-GAO	1.85 [g]	0.48	46

[a]. Smolders et al. (1994b); [b]. Zeng et al. (2003a); [c]. Oehmen et al. (2007b);
[d]. Oehment et al. (2006b).
[e]. C-mol/C-mol units.
[f]. Biomass fractions after executing a mathematical simulation for 50 days.
[g]. This study.

7.4.2. Effect of environmental and operational conditions

In the current study, the influence of different environmental and operational conditions on the PAO-GAO competition was evaluated by executing thirty-six different simulations using the integrated metabolic model. A summary of the bacterial population distributions observed at the end of the simulations according to the applied carbon sources, pH and temperature is displayed in Figure 7.6. It was considered that an organism was dominant, at the end of the 50-day mathematical simulations, if it comprised at least 60 % of the total bacterial population. If none of the microorganisms made up more than 60 %, it was assumed that the two highest fractions coexisted.

Temperature	100 % HAc			75-25 % HAc-HPr			50-50 % HAc-HPr			100 % HPr		
30 °C	Competi	Competi	Competi / PAO	PAO / Alpha	PAO	PAO	Alpha	Alpha / PAO	PAO	Alpha	Alpha	Alpha
20 °C	Competi / PAO	Competi / PAO	PAO	PAO	PAO	PAO	PAO	PAO	PAO	Alpha	Alpha / PAO	PAO
10 °C	PAO	PAO	PAO	PAO	PAO	PAO	PAO	PAO	PAO	PAO	PAO	PAO
pH	6.0	7.0	7.5	6.0	7.0	7.5	6.0	7.0	7.5	6.0	7.0	7.5

Figure 7.6. Summary of the bacterial population distributions showing the dominant or coexisting microorganisms as a function of the carbon source (acetate to propionate ratios), pH and temperature. White cells: *Accumulibacter* (PAO); black cells: *Competibacter* (GAO) or *Alphaproteobacteria-GAO* (GAO). Light grey tones indicate the coexistence of two microorganisms.

As observed in Figure 7.6, *Accumulibacter* dominated the PAO-GAO competition at 10 °C, suggesting that low temperature is detrimental for GAO. In accordance, Lopez-Vazquez *et al.* (2008c) observed that an enriched *Competibacter* culture was severely inhibited at 10 °C. Apparently, in that study, the anaerobic glycogen hydrolysis, which is assumed to be used by GAO as energy and carbon source for anaerobic substrate uptake, and the aerobic biomass production on PHA were the two metabolic pathways that limited the substrate uptake and growth of *Competibacter* at 10 °C. Lower sewage temperature (10 °C) appears to be beneficial for the EBPR process not because it favors the activity of PAO but due to the limiting effects that produces on the metabolism of GAO.

The sole presence of either HAc or HPr did not favor the dominance of PAO at moderate temperature (20 °C). With a pH 6.0-7.0, *Accumulibacter* and GAO tended to co-exist at 20 °C (Figure 7.6) due to their similar metabolisms that lead to comparable net biomass growth yields (Table 7.5). However, in practice, enriched PAO cultures have been cultivated in lab-scale studies using either HAc or HPr as sole carbon source (Smolders *et al.* 1995; Pijuan *et al.*, 2004; Oehmen *et al.*, 2005b; Lopez-Vazquez *et al.*, 2007b). In order to explain these discrepancies, 20 additional simulations were executed with different initial PAO-GAO fractions (90-10, 80-20, 70-30, 60-40 and 50-50 % *Accumulibacter* to *Competibacter* ratios) and influent P/HAc ratios (0.01, 0.04, 0.08 and 0.15 P-mol/C-mol) under standard conditions (HAc, 20 °C and pH 7.0). The competition was not influenced by the initial PAO-GAO ratios but by the low P/HAc ratio (0.01 P-mol/C-mol) (Figure 7.7). Due to the low influent phosphorus concentration, PAO were outcompeted because their intracellular poly-P pools were depleted, thus losing one of their main energy sources for substrate uptake. These observations are in accordance with previous reports (Liu *et al.*, 1997; Schuler and Jenkins, 2003). The substrate uptake and net biomass growth rates of PAO and GAO are so similar at standard conditions (20 °C, pH 7.0) that, unless they are affected by external factors (e.g. influent P/VFA ratio), the dominant microorganism present in the inoculum seem to be hardly outcompeted. Thus, in lab-scale systems operated at standard conditions, enriched PAO cultures are likely achieved because they are the dominant organisms in the inoculum and, furthermore, the applied operating conditions do not give GAO any opportunity to proliferate. Meanwhile, in order to get enriched GAO cultures, low P/VFA ratios are usually applied to outcompete PAO (Filipe *et al.*, 2001a; Zeng *et al.*, 2003a; Oehmen *et al.*, 2006b; Lopez-Vazquez *et al.*, 2008c). It is possible that the failure or inadequate performance of lab-scale EBPR reactors occurred (Cech *et al.*, 1993; Satoh *et al.*, 1994; Filipe *et al.*, 2001a) because the PAO and GAO fractions in the inoculum were similar and the operating conditions affected or limited the activity of PAO.

When both HAc and HPr were simultaneously supplied, *Accumulibacter* were the dominant organisms at 20 °C (Figure 7.6). Under these conditions, GAO are not able to simultaneously take up these two carbon sources as efficiently as *Accumulibacter*, which is consistent with the results of Chen *et al.* (2004).

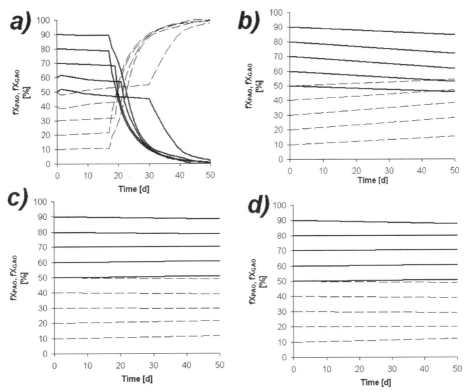

Figure 7.7. Effects of influent P/HAc ratios and initial biomass fractions (*Accumulibacter* and *Competibacter*) on the PAO-GAO competition: (a) 0.01 P-mol/C-mol P/HAc ratio, (b) 0.04 P-mol/C-mol P/HAc ratio, (c) 0.08 P-mol/C-mol P/HAc ratio, and (d) 0.15 P-mol/C-mol P/HAc ratio. *Accumulibacter*: bold black line; *Competibacter*: dashed grey line.

At high temperature (30 °C), and as consequence of their higher substrate uptake rates (Figure 7.1), GAO tended to dominate the competition when a single carbon source was provided (Figure 7.6). This is in agreement with other reports carried out at higher temperatures using HAc as the sole carbon source (Panswad *et al.*, 2003; Whang and Park, 2006). Nevertheless, there seem to be possibilities of achieving a good EBPR process performance at high temperature through supplying a 75-25 % HAc to HPr ratio and avoiding to use a pH lower than 7.0 (Figure 7.6). Under these conditions (Figure 7.4e), *Accumulibacter* had a total VFA uptake (36 %) slightly higher than those of *Competibacter* and *Alphaproteobacteria-GAO* (33 and 31 %, respectively). Moreover, the net growth yield of *Accumulibacter* was higher than that of *Competibacter* and similar to the net growth of *Alphaproteobacteria-GAO* (0.31, 0.26 and 0.32 C-mol active biomass/C-mol VFA, respectively). Thus, PAO were dominant because (i) *Competibacter* and *Alphaproteobacteria-GAO* were not able to take up both HAc and HPr with the same rate, and (ii) GAO did not have any considerable advantage regarding net biomass growth.

If similar HAc and HPr fractions (50-50 %) are supplied and a 7.0 pH is used at 30 °C, *Alphaproteobacteria-GAO* can proliferate due to their higher affinity for HPr, being able to take up about 40 % of the total supplied VFA. This is comparable to the total VFA taken up by *Accumulibacter*, which was approximately 43 %, resulting in the coexistence of these two microorganisms (Figures 4h and 6). Under these operational conditions (30 °C and 50-50 % HAc to HPr ratio) but at a higher pH (7.5) (Figure 7.4i), *Alphaproteobacteria-GAO* took up 22 % of the total VFA whereas *Accumulibacter* consumed around 77 %. Therefore, besides providing a proper HAc to HPr ratio, a pH higher than 7.0 (e.g. 7.5) appears to be also recommendable to maintain a suitable environment for PAO at high temperature (30 °C) if the HAc to HPr ratio entering the anaerobic zone is approximately 50-50 %.

7.4.3. Implications for lab- /full-scale EBPR systems

The dominance of PAO at 10 °C, which results from the inhibitory effects that GAO suffered at this temperature, helps to explain the generally stable operation of EBPR systems operated under cold weather conditions (van Veldhuizen *et al.*, 1999; Ybstebɸ *et al.*, 2000; Lopez-Vazquez *et al.*, 2008a). The results of the present study indicated that a mixture of HAc-HPr in the influent (in the order of 75-25 or 50-50 % HAc to HPr ratio), as well as a high pH (7.5), are beneficial for PAO at moderate temperature (20 °C). This is in agreement with previous full-scale observations where stable EBPR performance was observed with similar influent carbon source fractions (Meijer *et al.*, 2001; Lopez-Vazquez *et al.*, 2008a) and pH levels above 7.0 (Lopez-Vazquez *et al.*, 2008a).

In general, a relatively high temperature (30 °C) appeared to be more favorable for GAO. Accordingly, Gu *et al.* (2005) described the seasonal deterioration of the EBPR process performance as wastewater temperature increased during the summer. As observed in this study, a suitable influent HAc to HPr ratio (75-25 %) and pH not lower than 7.0 could be potentially useful to favor the metabolism of PAO at a higher temperature (i.e. 30 °C). Thereupon, optimizing the EBPR process performance at sewage temperatures higher than 20 °C (e.g. during summer, in warm regions or when treating warm industrial wastewater) may require a proper operation of the prefermenters to manipulate the influent VFA fractions, as reported elsewhere (Thomas *et al.*, 2003). Different strategies could also be evaluated towards the implementation of pH controls in equalizer, primary sedimentation or anaerobic tanks. Moreover, assuming that a higher temperature enhances the biological nitrification processes, an adequate operation of the internal recirculation flow-rates is necessary to favor the operation of well-defined anoxic and anaerobic stages, reducing the risk of nitrate and/or nitrite intrusion into the anaerobic stages. A good denitrification activity may also result in a higher pH in the anoxic phase, which, in combination with well-operated internal recirculation flows, could help to increase the pH in the anaerobic phase having a positive influence on the anaerobic metabolism of PAO.

In full-scale systems, the occurrence of *Alphaproteobacteria-GAO* (such as *Defluviicoccus vanus* related organisms) seems to be linked to industrial discharges (Burow *et al.*, 2007). According to the current integrated model, *Alphaproteobacteria-GAO* tend to proliferate at 30 °C when HPr is present in the influent (Figure 7.6). These observations may help to explain the appearance of *Alphaproteobacteria-GAO* at full-scale EBPR systems, assuming that industrial wastewaters are usually warmer and, from a carbon source perspective, more complex (potentially having a higher HPr fraction) than domestic sewage.

In the current research, it was assumed that PAO grown on HAc and HPr had the same temperature dependencies. A similar approach was followed to describe the pH and temperature dependencies of GAO cultivated with HAc and HPr. This assumption was made based on the successful application of EBPR mathematical models at full-scale systems regardless of the potential presence of different strains (Veldhuizen *et al.*, 1999; Brdjanovic *et al.*, 2000; Meijer *et al.*, 2001, 2002). Nevertheless, whenever other dependencies are defined (e.g. the temperature dependencies of *Accumulibacter* on HPr) they could be incorporated into the current integrated metabolic model, thereby improving its reliability, reach and scope.

Following a similar approach to Meijer *et al.* (2001), the implementation of the present model at full-scale systems could be an important tool towards the optimization of EBPR processes being able to predict, and therefore avoid, potential process upsets and deteriorations as consequence of the interaction between PAO and GAO.

7.5. Conclusions

The influence of certain environmental and operational conditions on the PAO-GAO competition were evaluated using an integrated metabolic model that incorporated their carbon source, temperature and pH dependencies. Independently of the carbon source or pH, PAO were the dominant microorganisms at low temperature (10 °C) since the metabolism of GAO was inhibited at 10 °C. At moderate temperature (20 °C), the simultaneous presence of acetate and propionate as carbon sources (e.g. 75-25 or 50-50 % acetate to propionate ratios) favored the growth of PAO over GAO, regardless of the applied pH. In contrast, when either acetate or propionate is supplied as sole carbon source, PAO is only favored over GAO when a high pH (7.5) is applied. In general, GAO (*Competibacter* and *Alphaproteobacteria-GAO*) tended to proliferate at higher temperature (30 °C). Nevertheless, an adequate acetate to propionate ratio (75 – 25 %) and a pH not lower than 7.0 could be used as potential tools to suppress the proliferation of GAO at high temperature (30 °C). Metabolic modeling can be a highly useful means of describing key factors affecting the complex interactions between different microorganisms relevant in EBPR systems, and in predicting their resulting population dynamics. Incorporating the main factors affecting the PAO-GAO competition that are described in this study into future modeling endeavors focused on the optimization of full-scale EBPR plants would likely facilitate improved process efficiency and robustness.

Acknowledgements

Carlos Lopez-Vazquez would like to acknowledge the Mexican National Council for Science and Technology (CONACYT) as well as the Autonomous University of the State of Mexico (Toluca, Mexico) for doctoral grant. Adrian Oehmen would like to acknowledge the Fundação para a Ciência e Tecnologia for postdoctoral grant SFRH/BPD/41486/2007. Zhiguo Yuan would like to acknowledge the financial support provided by the Natural Science Foundation of China via project 50628808 and the Australian Research Council (ARC) via project DP0342658.

References

Beun JJ, Paletta F, van Loosdrecht MCM, Heijnen JJ (2000a) Stoichiometry and kinetics of poly-β-hydroxybutyrate metabolism in aerobic, slow growing, activated sludge culture. Biotechnol Bioeng 67(4), 379-389.

Beun JJ, Verhoef EV, van Loosdrecht MCM, Heijnen JJ (2000b) Stoichiometry and kinetics of poly-β-hydroxybutyrate metabolism under denitrifying conditions in activated sludge cultures. Biotechnol Bioeng 68(5), 496-507.

Brdjanovic D, van Loosdrecht MCM, Hooijmans CM, Alaerts GJ, Heijnen JJ (1997) Temperature effects on physiology of biological phosphorus removal. ASCE J Environ Eng 123(2), 144-154.

Brdjanovic D, Logemann S, van Loosdrecht MCM, Hooijmans CM, Alaerts GJ, Heijnen JJ (1998) Influence of temperature on biological phosphorus removal: process and molecular ecological studies. Water Res 32(4), 1035-1048.

Brdjanovic D, van Loosdrecht MCM, Versteeg P, Hooijmans CM, Alaerts GJ, Heijnen JJ (2000) Modeling COD, N and P removal in a full-scale wwtp Haarlem Waarderpolder. Water Res., 34(3), 846-858.

Burow LC, Kong Y, Nielsen JL, Blackall LL, Nielsen PH (2007) Abundance and ecophysiology of *Defluviicoccus spp.*, glycogen-accumulating organisms in full-scale wastewater treatment processes. Microbiology-SGM 153, 178-185.

Cech JS, Hartman P (1993) Competition between phosphate and polysaccharide accumulating bacteria in enhanced biological phosphorus removal systems. Water Res 27, 1219-1225.

Chen Y, Randall AA, McCue T (2004) The efficiency of enhanced biological phosphorus removal from real wastewater affected by different ratios of acetic and propionic acid. Water Res 38(1), 27-36.

Dai Y, Yuan Z, Wang X, Oehmen A, Keller J (2007) Anaerobic metabolism of *Defluviicoccus vanus* related glycogen accumulating organisms (GAOs) with acetate and propionate as carbon sources. Wat Res 41(9), 1885-1896.

Dias JML, Oehmen A, Serafim LS, Lemos PC, Reis MAM, Olivera R (2008) Metabolic modelling of polyhydroxyalkanoate copolymers production by mixed microbial cultures. BMC Systems Bio (accepted).

Dircks K, Beun JJ, van Loosdrecht MCM, Heijnen JJ, Henze M (2001) Glycogen metabolism in aerobic mixed cultures. Biotechnol Bioeng 73(2), 85-94.

Filipe CDM, Daigger GT, Grady Jr CPL (2001a) A metabolic model for acetate uptake under anaerobic conditions by glycogen-accumulating organisms: stoichiometry, kinetics and effect of pH. Biotechnol Bioeng 76(1):17-31.

Filipe CDM, Daigger GT, Grady Jr CPL (2001b) Effects of pH on the aerobic metabolism of phosphate-accumulating organisms and glycogen-accumulating organisms. Water Environ Res 73(2), 213-222.

Filipe CDM, Daigger GT, Grady Jr CPL (2001c) pH as a key factor in the competition between glycogen-accumulating organisms and phosphorus-accumulating organisms. Water Environ Res 73(2), 223-232.

Filipe CDM, Daigger GT, Grady Jr CPL (2001d) Stoichiometry and kinetics of acetate uptake under anaerobic conditions by an enriched culture of phosphorus-accumulating organisms at different pH. Biotechnol Bioeng 76(1), 32-43.

Gu AZ, Saunders AM, Neethling JB, Stensel HD, Blackall L (2005) In: WEF (Ed.) Investigation of PAOs and GAOs and their effects on EBPR performance at full-scale wastewater treatment plants in US, October 29–November 2, WEFTEC, Washington, DC, USA.

Kuba T, Murnleiter E, van Loosdrecht MCM, Heijnen JJ (1996) A metabolic model for biological phosphorus removal by denitrifying organisms. Bioetchnol Bioeng 52(6), 685-695.

Liu WT, Nakamura K, Matsuo T, Mino T (1997) Internal energy-based competition between polyphosphate- and glycogen-accumulating bacteria in biological phosphorus removal reactors-effect of P/C feeding ratio. Water Res 31(6), 1430-1438.

Lopez-Vazquez CM, Hooijmans CM, Brdjanovic D, Gijzen HJ, van Loosdrecht, MCM (2007a) A practical method for the quantification of phosphorus and glycogen-accumulating organisms in activated sludge systems. Water Environ Res 79(13), 2487-2498.

Lopez-Vazquez CM, Song YI, Hooijmans CM, Brdjanovic D, Moussa MS, Gijzen HJ, van Loosdrecht MCM (2007b) Short-term temperature effects on the anaerobic metabolism of glycogen accumulating organisms. Biotech Bioeng 97(3), 483-495.

Lopez-Vazquez CM, Hooijmans CM, Brdjanovic D, Gijzen HJ, van Loosdrecht MCM (2008a) Factors affecting the microbial populations at full-scale Enhanced Biological Phosphorus Removal (EBPR) wastewater treatment plants in The Netherlands. Water Res 42(10-11), 2349-2360.

Lopez-Vazquez CM, Song YI, Hooijmans CM, Brdjanovic D, Moussa MS, Gijzen HJ, van Loosdrecht MCM (2008b) Temperature effects on the aerobic metabolism of glycogen accumulating organisms. Biotech Bioeng 101(2):295-306.

Lopez-Vazquez CM, Hooijmans CM, Brdjanovic D, Gijzen HJ, van Loosdrecht MCM (2008c) Long-term temperature effects on the metabolism of glycogen-accumulating organisms (submitted).

Manga J, Ferrer J, Garcia-Usach F, Seco A (2001) A modification to the Activated Sludge Model No. 2. Water Sci Technol 43(11), 161-171.

Meijer SCF, van Loosdrecht MCM, Heijnen JJ (2001) Metabolic modelling of full-scale biological nitrogen and phosphorus removing WWTP's. Water Res 35(11), 2711-2723.

Meijer SCF, van Loosdrecht MCM, Heijnen JJ (2002) Modelling the start-up of a full-scale biological nitrogen and phosphorus removing WWTP's. Water Res 36(11), 4667-4682.

Murnleitner E, Kuba T, van Loosdrecht MCM, Heijnen JJ (1997) An integrated metabolic model for the aerobic and denitrifying biological phosphorus removal. Biotechnol Bioeng 54(5), 434-450.

Oehmen A (2004) The competition between polyphosphate accumulating organisms and glycogen accumulating organisms in the enhanced biological phosphorus removal process. PhD thesis. The University of Queensland. Brisbane, Australia.

Oehmen A, Yuan Z, Blackall LL, Keller J (2005a) Comparison of acetate and propionate uptake by polyphosphate accumulating organisms and glycogen accumulating organisms. Biotechnol Bioeng 91(2), 162-168.

Oehmen A, Zeng RJ, Yuan Z, Keller J (2005b) Anaerobic metabolism of propionate by polyphosphate-accumulating organisms in enhanced biological phosphorus removal systems. Biotechnol Bioeng 91(1), 43-53.

Oehmen A, Vives MT, Lu H, Yuan Z, Keller J (2005c) The effect of pH on the competition between polyphosphate-accumulating organisms and glycogen-accumulating organisms. Water Res 39(15), 3727-3737.

Oehmen A, Saunders AM, Vives MT, Yuan Z, Keller J (2006a) Competition between polyphosphate and glycogen accumulating organisms in enhanced biological phosphorus removal systems with acetate and propionate as carbon sources. J Biotechnol 123(1), 22-32.

Oehmen A, Zeng RJ, Saunders AM, Blackall LL, Keller J, Yuan Z (2006b) Anaerobic and aerobic metabolism of glycogen accumulating organisms selected with propionate as the sole carbon source. Microbiology 152(9), 2767-2778.

Oehmen A, Lemos PC, Carvalho G, Yuan Z, Keller J, Blackall LL, Reis MAM (2007a) Advances in enhanced biological phosphorus removal: from micro to macro scale. Water Res 41(11), 2271-2300.

Oehmen A, Zeng RJ, Keller J, Yuan Z (2007b) Modeling the aerobic metabolism of polyphosphate-accumulating organisms enriched with propionate as a carbon source. Water Environ Res 79(13), 2477-2486.

Panswad T, Doungchai A, Anotai J (2003) Temperature effect on microbial community of enhanced biological phosphorus removal system. Water Res 37, 409-415.

Pijuan M, Saunders AM, Guisasola A, Baeza JA, Casas C, Blackall LL (2004) Enhanced biological phosphorus removal in a sequencing batch reactor using propionate as the sole carbon source. Biotechnol Bioeng 85(1), 56-67.

Reichert P (1994) AQUASIM - a tool for simulation and data analysis of aquatic systems. Water Sci Technol 30(2), 21-30.

Salem S, Berends DHJG, van Loosdrecht MCM, Heijnen J (2002) Model-based evaluation of a new upgrading concept for nitrogen removal. Water Sci Technol 45(6), 169-176.

Satoh H, Mino T, Matsuo T (1994) Deterioration of enhanced biological phosphorus removal by the domination of microorganisms without polyphosphate accumulation. Water Sci Technol 30(6), 203-211.

Saunders AM, Oehmen A, Blackall LL, Yuan Z, Keller J (2003) The effect of GAO (glycogen accumulating organisms) on anaerobic carbon requirements in full-scale Australian EBPR (enhanced biological phosphorus removal) plants. Water Sci Technol 47(11), 37-43.

Schuler AJ, Jenkins D (2003) Enhanced biological phosphorus removal from wastewater by biomass with different phosphorus contents, Part 1: Experimental results and comparison with metabolic models. Water Environ Res 75(6), 485-498.

Smolders GJF, van der Meij J, van Loosdrecht MCM, Heijnen JJ (1994a) Model of the anaerobic metabolism of the biological phosphorus removal process: stoichiometry and pH influence. Biotechnol Bioeng 43(6), 461-470.

Smolders GJF, van der Meij J, van Loosdrecht MCM, Heijnen JJ (1994b) Stoichiometric model of the aerobic metabolism of the biological phosphorus removal process. Biotechnol Bioeng 44(7), 837-848.

Smolders GJF, van der Meij J, van Loosdrecht MCM, Heijnen JJ (1995) A structured metabolic model for anaerobic and aerobic stoichiometry and kinetics of the biological phosphorus removal process. Biotechnol Bioeng 47(3), 277-287.

Thomas M, Wright P, Blackall L, Urbain V, Keller J (2003) Optimisation of Noosa BNR plant to improve performance and reduce operating costs. Water Sci Technol 47 (12), 141–148.

van Veldhuizen HM, van Loosdrecht MCM, Heijnen JJ (1999) Modelling biological phosphorus and nitrogen removal in a full scale activated sludge process. Water Res 33(16), 3459-3468.

Whang LM, Park JK (2006) Competition between polyphosphate- and glycogen-accumulating organisms in enhanced biological phosphorus removal systems: effect of temperature and sludge age. Water Environ Res 78(1), 4-11.

Whang LM, Filipe CDM, Park JK (2007) Model-based evaluation of competition between polyphosphate- and glycogen-accumulating organisms. Water Res 41(6), 1312-1324.

Yagci N, Artan N, Cogkor EU, Randall C, Orhon D (2003) Metabolic model for acetate uptake by a mixed culture of phosphate- and glycogen-accumulating organisms under anaerobic conditions. Biotechnol Bioeng 84(3), 359-373.

Yagci N, Insel G, Orhon D (2004) Modelling and calibration of phosphate and glycogen accumulating organism competition for acetate uptake in a sequencing batch reactor. Water Sci Technol 50(6), 241-250.

Ydstebф L, Bilstad T, Barnard J (2000) Experience with biological nutrient removal at low temperatures. Water Environ Res 72(4), 444-454.

Zeng RJ, van Loosdrecht MCM, Yuan Z, Keller J (2003a) Metabolic model for glycogen-accumulating organisms in anaerobic/aerobic activated sludge systems. Biotechnol Bioeng 81(1), 92-105.

Zeng RJ, Yuan Z, Keller J (2003b) Model-based analysis of anaerobic acetate uptake by a mixed culture of polyphosphate-accumulating and glycogen-accumulating organisms. Biotechnol Bioeng 83(3), 293-302.

Appendix 7.1.

Stoichiometric matrix and components for *Accumulibacter, Competibacter* and

| | Components | \multicolumn{6}{c}{Components} | | | | | |
| | | 1 | 2 | 3 | 4 | 5 | 6 |
		S_{O2}	S_{HAc}	$S_{H\,Pr}$	S_{PO4}	X_{PAO}	X_{PHA}^{PAO}
	Accumulibacter						
1	Anaerobic acetate uptake		-1		$Y_{PO4,HAc}^{PAO}$		$Y_{PHA,HAc}^{PAO}$
2	Anaerobic propionate uptake			-1	$Y_{PO4,H\,Pr}^{PAO}$		$Y_{PHA,H\,Pr}^{PAO}$
3	Anaerobic maintenance				1		
4	Aerobic PHA degradation	$-i_{O2,PHA}^{PAO}$			$-i_{PO4,X}^{PAO}$	$Y_{S,X}^{PAO}$	-1
5	Aerobic glycogen production	$i_{O2,GLY}^{PAO}$			$i_{PO4,GLY}^{PAO}$	$-i_{PAO,GLY}$	
6	Aerobic Poly-P formation	$i_{O2,PP}^{PAO}$			$-i_{PO4,PP}^{PAO}$	$-i_{PAO,PP}$	
7	Aerobic maintenance	$i_{O2,mAER}^{PAO}$			$i_{PO4,mAER}^{PAO}$	$-Y_{S,X}^{PAO}$	
	Competibacter						
8	Anaerobic acetate uptake		-1				
9	Anaerobic propionate uptake			-1			
10	Anaerobic maintenance						
11	Aerobic PHA degradation	$-i_{O2,PHA}^{Competi}$			$-i_{PO4,X}^{Competi}$		
12	Aerobic glycogen production	$i_{O2,GLY}^{Competi}$			$i_{PO4,GLY}^{Competi}$		
13	Aerobic maintenance	$i_{O2,mAER}^{Competi}$			$i_{PO4,mAER}^{Competi}$		
	Alphaproteobacteria						
14	Anaerobic acetate uptake		-1				
15	Anaerobic propionate uptake			-1			
16	Anaerobic maintenance	$-i_{O2,PHA}^{Alpha}$			$-i_{PO4,X}^{Alpha}$		
17	Aerobic PHA degradation	$i_{O2,GLY}^{Alpha}$			$i_{PO4,GLY}^{Alpha}$		
18	Aerobic PHA degradation	$i_{O2,mAER}^{Alpha}$			$i_{PO4,mAER}^{Alpha}$		
19	Aerobic glycogen production						

Alphaproteobacteria-GAO.

	Components						
7	**8**	**9**	**10**	**11**	**12**	**13**	**14**
X_{GLY}^{PAO}	X_{PP}	$X_{Competi}$	$X_{PHA}^{Competi}$	$X_{GLY}^{Competi}$	X_{Alpha}	X_{PHA}^{Alpha}	X_{GLY}^{Alpha}
$-Y_{GLY,HAc}^{PAO}$	$-Y_{PO4,HAc}^{PAO}$						
$-Y_{GLY,HPr}^{PAO}$	$-Y_{PO4,HPr}^{PAO}$						
	-1						
1							
	1						
			$Y_{PHA,HAc}^{Competi}$	$-Y_{GLY,HAc}^{Competi}$			
			$Y_{PHA,HPr}^{Competi}$	$-Y_{GLY,HPr}^{Competi}$			
			$Y_{PHA,mANA}^{Competi}$	$-Y_{GLY,mANA}^{Competi}$			
		$Y_{S,X}^{Competi}$	-1				
		$-i_{Competi,GLY}$		1			
		$-Y_{S,X}^{Competi}$					
						$Y_{PHA,HAc}^{Alpha}$	$-Y_{GLY,HAc}^{Alpha}$
						$Y_{PHA,HPr}^{Alpha}$	$-Y_{GLY,HPr}^{Alpha}$
						$Y_{PHA,mANA}^{Alpha}$	$-Y_{GLY,mANA}^{Alpha}$
					$Y_{S,X}^{Alpha}$	-1	
					$-i_{Alpha,GLY}$		1
					$-Y_{S,X}^{Alpha}$		

where:

$$i_{O2,PHA}^{PAO} = Y_{S,X}^{PAO} / Y_{O,X}^{PAO}$$

$$i_{O2,GLY}^{PAO} = Y_{S,X}^{PAO} / (Y_{O,X}^{PAO} \cdot Y_{S,GLY}^{PAO}) - (1/Y_{O,GLY}^{PAO})$$

$$i_{O2,PP}^{PAO} = Y_{S,X}^{PAO} / (Y_{O,X}^{PAO} \cdot Y_{S,PP}^{PAO}) - (1/Y_{O,PP}^{PAO})$$

$$i_{O2,mAER}^{PAO} = (Y_{S,X}^{PAO} / Y_{O,X}^{PAO}) - 1$$

$$i_{PO4,X}^{PAO} = i_{BM,P} \cdot Y_{S,X}^{PAO}$$

$$i_{PO4,GLY}^{PAO} = i_{BM,P} \cdot Y_{S,X}^{PAO} / Y_{S,GLY}^{PAO}$$

$$i_{PO4,PP}^{PAO} = (i_{BM,P} \cdot Y_{S,X}^{PAO} / Y_{S,PP}^{PAO}) - 1$$

$$i_{PO4,mAER}^{PAO} = i_{BM,P} \cdot Y_{S,X}^{PAO}$$

$$i_{PAO,GLY} = -Y_{S,X}^{PAO} / Y_{S,GLY}^{PAO}$$

$$i_{PAO,PP} = -Y_{S,X}^{PAO} / Y_{S,PP}^{PAO}$$

$$i_{O2,PHA}^{Competi} = Y_{S,X}^{Competi} / Y_{O,X}^{Competi}$$

$$i_{O2,GLY}^{Competi} = Y_{S,X}^{Competi} / (Y_{O,X}^{Competi} \cdot Y_{S,GLY}^{Competi}) - (1/Y_{O,GLY}^{Competi})$$

$$i_{O2,mAER}^{Competi} = (Y_{S,X}^{Competi} / Y_{O,X}^{Competi}) - 1$$

$$i_{PO4,X}^{Competi} = i_{BM,P} \cdot Y_{S,X}^{Competi}$$

$$i_{PO4,GLY}^{Competi} = i_{BM,P} \cdot Y_{S,X}^{Competi} / Y_{S,GLY}^{Competi}$$

$$i_{PO4,mAER}^{Competi} = i_{BM,P} \cdot Y_{S,X}^{Competi}$$

$$i_{Competi,GLY} = -Y_{S,X}^{Competi} / Y_{S,GLY}^{Competi}$$

$$i_{O2,PHA}^{Alpha} = Y_{S,X}^{Alpha} / Y_{O,X}^{Alpha}$$

$$i_{O2,GLY}^{Alpha} = Y_{S,X}^{Alpha} / (Y_{O,X}^{Alpha} \cdot Y_{S,GLY}^{Alpha}) - (1/Y_{O,GLY}^{Alpha})$$

$$i_{O2,mAER}^{Alpha} = (Y_{S,X}^{Alpha} / Y_{O,X}^{Alpha}) - 1$$

$$i_{PO4,X}^{Alpha} = i_{BM,P} \cdot Y_{S,X}^{Alpha}$$

$$i_{PO4,GLY}^{Alpha} = i_{BM,P} \cdot Y_{S,X}^{Alpha} / Y_{S,GLY}^{Alpha}$$

$$i_{PO4,mAER}^{Alpha} = i_{BM,P} \cdot Y_{S,X}^{Alpha}$$

$$i_{Alpha,GLY} = -Y_{S,X}^{Alpha} / Y_{S,GLY}^{Alpha}$$

Appendix 7.2.

Anaerobic stoichiometric parameters of *Accumulibacter.*

Parameter	Value	Units	Description	Source
$Y_{PHA,HAc}^{PAO}$	$(4/3)X_{PHB}^{PAO}$	C mmol	PHA stored per 1 C-mmol acetate taken up	Smolders et al. (1994a)
$Y_{GLY,HAc}^{PAO}$	$1/2$	C mmol	Glycogen consumed per 1 C-mmol acetate taken up	Smolders et al. (1994a)
$Y_{PO4,HAc}^{PAO}$	$1/2 + \alpha_{HAc}^{PAO}$	P mmol	Poly-P consumed per 1 C-mmol acetate taken up	Smolders et al. (1994a)
α_{HAc}^{PAO}	$0.16 \cdot pH - 0.7985$	ATP mmol	ATP necessary to transport 1 C-mmol acetate through cell membrane	Filipe et al. (2001d)
$Y_{PHA,HPr}^{PAO}$	*If pH $\leq$ 7.0 then:* $0.56 \cdot X_{PHV}^{PAO} + 0.67 \cdot X_{PH2MV}^{PAO} = 1.23 \cdot X_{PHA}^{PAO}$ *If pH > 7.0 then:* $\left. \begin{array}{l} (2.0217 - 0.21 \cdot pH)X_{PHV}^{PAO} \\ + \\ (1.1716 - 0.07 \cdot pH)X_{PH2MV}^{PAO} \end{array} \right\} X_{PHA}^{PAO}$	C mmol	PHA produced per 1 C-mmol propionate taken up	Oehmen et al. (2005b), Oehmen et al. (2005c)
$Y_{GLY,HPr}^{PAO}$	*If pH $\leq$ 7.0 then:* $1/3$ *If pH > 7.0 then:* $1.73 - 0.2 \cdot pH$	C mmol	Glycogen consumed per 1 C-mmol propionate stored	Oehmen et al. (2005b), Oehmen et al. (2005c)
$Y_{PO4,HPr}^{PAO}$	*If pH $\leq$ 7.5 then:* $2/9 - \alpha_{HPr}^{PAO}$ *If pH > 7.5 then:* $0.20 \cdot pH - 1.1$	P mmol	Poly-P consumed per 1 C-mmol propionate stored	Oehmen et al. (2005b), Oehmen et al. (2005c)
α_{HPr}^{PAO}	0.18	ATP mmol	ATP necessary to transport 1 C-mmol acetate through cell membrane	Oehmen et al. (2005b)

Anaerobic stoichiometric parameters of *Competibacter.*

Parameter	Value	Units	Description	Source
$Y_{PHA,HAc}^{Competi}$	$\dfrac{\left(\frac{9}{6}+\frac{2}{3}\alpha\right)^2}{\left(\frac{5}{3}+\frac{4}{3}\alpha\right)}X_{PHB}^{Competi}+$ $+\dfrac{2.5\left(\frac{9}{6}+\frac{2}{3}\alpha\right)\left(\frac{1}{6}+\frac{2}{3}\alpha\right)}{\left(\frac{5}{3}+\frac{4}{3}\alpha\right)}X_{PHV}^{Competi}+$ $+\dfrac{1.5\left(\frac{1}{6}+\frac{2}{3}\alpha\right)^2}{\left(\frac{5}{3}+\frac{4}{3}\alpha\right)}X_{PH2MV}^{Competi}$ $\Bigg\}\,X_{PHA}^{Competi}$	C mmol	PHA stored per 1 C-mmol acetate taken up	Zeng et al. (2003b)
$Y_{GLY,HAc}^{Competi}$	$1+2\cdot\alpha_{HAc}^{Competi}$	C mmol	Glycogen consumed per 1 C-mmol acetate taken up	Filipe et al. (2001a)
$\alpha_{HAc}^{Competi}$	$0.057\cdot pH-0.34$	ATP mmol	ATP necessary to transport 1 C-mmol acetate through cell membrane	Filipe et al. (2001a)
$Y_{PHA,HPr}^{Competi}$	$0.05\cdot X_{PHB}^{Competi}+$ $+0.70\cdot X_{PHV}^{Competi}+$ $+0.75\cdot X_{PH2MV}^{Competi}$ $\Bigg\}\,X_{PHA}^{Competi}$	C mmol	PHA produced per 1 C-mmol propionate taken up	Oehmen et al. (2006b)
$Y_{GLY,HPr}^{Competi}$	$0.67\cdot X_{GLY}^{Competi}$	C mmol	Glycogen consumed per 1 C-mmol propionate stored	Oehmen et al. (2006b)

Anaerobic stoichiometric parameters of *Alphaproteobacteria-GAO*.

Parameter	Value	Units	Description	Source
$Y_{PHA,HAc}^{Alpha}$	$\left. \begin{array}{l} \dfrac{\left(\dfrac{9}{6}+\dfrac{2}{3}\alpha\right)^2}{\left(\dfrac{5}{3}+\dfrac{4}{3}\alpha\right)}X_{PHB}^{Alpha}+ \\[2em] +\dfrac{2.5\left(\dfrac{9}{6}+\dfrac{2}{3}\alpha\right)\left(\dfrac{1}{6}+\dfrac{2}{3}\alpha\right)}{\left(\dfrac{5}{3}+\dfrac{4}{3}\alpha\right)}X_{PHV}^{Alpha}+ \\[2em] +\dfrac{1.5\left(\dfrac{1}{6}+\dfrac{2}{3}\alpha\right)^2}{\left(\dfrac{5}{3}+\dfrac{4}{3}\alpha\right)}X_{PH2MV}^{Alpha} \end{array} \right\} X_{PHA}^{Alpha}$	C mmol	PHA stored per 1 C-mmol acetate taken up	Zeng et al. (2003b)
$Y_{GLY,HAc}^{Alpha}$	$1+2\cdot\alpha_{HAc}^{Alpha}$	C mmol	Glycogen consumed per 1 C-mmol acetate taken up	Filipe et al. (2001a)
α_{HAc}^{Alpha}	$0.057\cdot pH-0.34$	ATP mmol	ATP necessary to transport 1 C-mmol acetate through cell membrane	Filipe et al. (2001a)
$Y_{PHA,H\,Pr}^{Alpha}$	$\left. \begin{array}{l} \dfrac{2\left(\dfrac{1}{2}+\alpha_{GAO}\right)^2}{3\left(\dfrac{5}{3}+2\alpha_{GAO}\right)}X_{PHB}^{Alpha}+ \\[2em] +\dfrac{5\left(\dfrac{1}{2}+\alpha_{GAO}\right)\left(\dfrac{7}{6}+\alpha_{GAO}\right)}{3\left(\dfrac{5}{3}+2\alpha_{GAO}\right)}X_{PHV}^{Alpha}+ \\[2em] +\dfrac{1.5\left(\dfrac{1}{6}+\dfrac{2}{3}\alpha\right)^2}{\left(\dfrac{5}{3}+\dfrac{4}{3}\alpha\right)}X_{PH2MV}^{Alpha} \end{array} \right\} X_{PHA}^{Alpha}$	C mmol	PHA stored per 1 C-mmol propionate taken up	Oehmen et al. (2006b)
$Y_{GLY,H\,Pr}^{Alpha}$	$2/3+2\cdot\alpha_{H\,Pr}^{Alpha}$	C mmol	Glycogen consumed per 1 C-mmol pro pionate taken up	Oehmen et al. (2006b)
$\alpha_{H\,Pr}^{Alpha}$	0	ATP mmol	ATP necessary to transport 1 C-mmol acetate through cell membrane	Oehmen et al. (2006b)

Appendix 7.3

Aerobic stoichiometric parameters for *Accumulibacter, Competibacter* and

Parameter	*Accumulibacter*	*Competibacter/Alphaproteobacteria-GAO*
$Y_{S,X}$	$\dfrac{250(106\lambda+127\beta)(6\lambda+27\lambda\delta+8\beta+30\beta\delta)}{(201930\lambda+318000K_1\lambda+678771\lambda\delta+813435\beta\delta+269240\beta+381000K_2\beta)}$	$\dfrac{250(106\lambda+127\beta)(6\lambda+27\lambda\delta+8\beta+30\beta\delta)}{(201930\lambda+318000K_1\lambda+678771\lambda\delta+813435\beta\delta+269240\beta+381000K_2\beta)}$
$Y_{S,PP}$	$\dfrac{\varepsilon(6\lambda+27\lambda\delta+8\beta+30\beta\delta)}{12(\varepsilon+\delta)}$	-
$Y_{S,GLY}$	$\dfrac{(3\lambda+4\beta)(6\lambda+27\lambda\delta+8\beta+30\beta\delta)}{24(2\lambda+3\lambda\delta+2\beta+4\beta\delta)}$	$\dfrac{(3\lambda+4\beta)(6\lambda+27\lambda\delta+8\beta+30\beta\delta)}{24(2\lambda+3\lambda\delta+2\beta+4\beta\delta)}$
m_S	$\dfrac{12m_{ATP}^{AER}}{6\lambda+27\lambda\delta+8\beta+30\beta\delta}$ *where* $m_{ATP}^{AER}=0.019ATP-mol/(C-mol-h)$	$\dfrac{12m_{ATP}^{AER}}{6\lambda+27\lambda\delta+8\beta+30\beta\delta}$ *where* $m_{ATP}^{AER}=0.019ATP-mol/(C-mol-h)$
$Y_{O,X}$	$\left(\dfrac{\mathrm{Re}doxPHA}{4}\cdot\dfrac{1}{Y_{S,X}^{PAO}}-\dfrac{\mathrm{Re}doxBM}{4}\right)^{-1}$	$\left(\dfrac{\mathrm{Re}doxPHA}{4}\cdot\dfrac{1}{Y_{S,X}^{Competi}}-\dfrac{\mathrm{Re}doxBM}{4}\right)^{-1}$
$Y_{O,PP}$	$\left(\dfrac{\mathrm{Re}doxPHA}{4}\cdot\dfrac{1}{Y_{S,PP}^{PAO}}\right)^{-1}$	-
$Y_{O,GLY}$	$\left(\dfrac{\mathrm{Re}doxPHA}{4}\cdot\dfrac{1}{Y_{S,GLY}^{PAO}}-1\right)^{-1}$	$\left(\dfrac{\mathrm{Re}doxPHA}{4}\cdot\dfrac{1}{Y_{S,GLY}^{Competi}}-1\right)^{-1}$
RedoxPHA	$4.5a+4.8b+5c$	$4.5a+4.8b+5c$
λ	$a+(2b/5)$	$a+(2b/5)$
β	$c+(3b/5)$	$c+(3b/5)$
a	$X_{PHB}^{PAO}/X_{PHA}^{PAO}$	$X_{PHB}^{Competi}/X_{PHA}^{Competi}$; $X_{PHB}^{Alpha}/X_{PHA}^{Alpha}$
b	$X_{PHV}^{PAO}/X_{PHA}^{PAO}$	$X_{PHV}^{Competi}/X_{PHA}^{Competi}$; $X_{PHV}^{Alpha}/X_{PHA}^{Alpha}$
c	$X_{PH2MV}^{PAO}/X_{PHA}^{PAO}$	$X_{PH2MV}^{Competi}/X_{PHA}^{Competi}$; $X_{PH2MV}^{Alpha}/X_{PHA}^{Alpha}$
RedoxBM	$4+i_{BM,H}-2\cdot i_{BM,O}-3\cdot i_{BM,N}+5\cdot i_{BM,P}$	$4+i_{BM,H}-2\cdot i_{BM,O}-3\cdot i_{BM,N}+5\cdot i_{BM,P}$
$i_{BM,H}$	1.84	1.84
$i_{BM,O}$	0.50	0.50
$i_{BM,N}$	0.19	0.19
$i_{BM,P}$	0.015	0.015

Alphaproteobacteria-GAO (Smolders et al. 1994b; Zeng et al. 2003a).

Units	Description
C-mmol /C-mmol	Maximum biomass growth yield on PHA
P-mmol/C-mmol	Maximum yield of poly-P stored on PHA
C-mmol/C-mmol	Maximum yield of glycogen stored on PHA
C-mmol/(C-mmol - h)	Aerobic maintenance coefficient on PHA
C-mmol/O_2-mmol	Oxygen consumed per biomass growth
P-mmol/O_2-mmol	Oxygen consumed per poly-P uptake
C-mmol/O_2-mmol	Oxygen consumed per glycogen produced
Number of electrons per C-mol	PHA degree of reduction
C-mmol/C-mmol	Percentage of Acetyl-CoA* in PHA
C-mmol/C-mmol	Percentage of Propionyl-CoA* in PHA
C-mmol/C-mmol	PHB fraction in PHA
C-mmol/C-mmol	PHV fraction in PHA
C-mmol/C-mmol	PH2MV fraction in PHA
Number of electrons per C-mol	Biomass degree of reduction
H-mol/C-mol	Hydrogen content in biomass
O-mol/C-mol	Oxygen content in biomass
N-mol/C-mol	Nitrogen content in biomass
P-mol/C-mol	Phosphorus content in biomass

Appendix 7.4

Aerobic metabolic parameters for *Accumulibacter*, *Competibacter* and *Alphaproteobacteria-GAO*.

Microorganisms	Parameter			
	δ	K_1	K_2	ε
	ATP-mol/NADH-mol	ATP-mol/C-mol	ATP-mol/C-mol	P-mol/NADH$_2$-mol
	ATP produced per NADH$_2$ oxidized (P/O ratio)	ATP needed for biomass synthesis from Acetyl-CoA*	ATP needed for biomass synthesis from Propionyl-CoA*	Phosphate transport coefficient
	Smolders et al. (1994b), This study	Smolders et al. (1994b), Zeng et al. (2003a)	Zeng et al. (2003a)	Smolders et al. (1994b)
Accumulibacter	1.85	1.7	1.38	7
Competibacter	1.85	1.7	1.38	-
Alphaproteobacteria GAO	1.85	1.7	1.38	-

Appendix 7.5.

Kinetic parameters for *Accumulibacter.*

	Process		Parameter
1,2	Maximum anaerobic acetate and propionate uptake rate	$q_{PAO,VFA}^{MAX}$	*If* $T \leq 20\,^oC$ *then* $$q_{PAO,VFA,20}^{MAX} \cdot 1.095^{(T-20)}$$ *If* $20 < T \leq 30\,^oC$ *then* $$q_{PAO,VFA,20}^{MAX}$$ *where:* $$q_{PAO,VFA,20}^{MAX} = q_{PAO,HAc,20}^{MAX} = q_{PAO,H\,Pr,20}^{MAX} = 0.20$$
3	Anaerobic maintenance rate	m_{PAO}^{ANA}	$$m_{ATP,PAO,20}^{ANA} \cdot 1.096^{(T-20)}$$ *where:* $$m_{ATP,PAO,20}^{ANA} = 2.35x10^{-3}\,ATP - mol/(C - mol - h)$$
4	Aerobic PHA degradation rate	k_{PHA}^{PAO}	*If* $T \leq 20\,^oC$ *then* $$\left[(k_{PHA,HAc,20}^{PAO} - k_{PHA,H\,Pr,20}^{PAO}) \cdot (f_{HAc,INF}) + k_{PHA,H\,Pr,20}^{PAO} \right] \cdot 1.129^{(T-20)}$$ *If* $20 < T \leq 30\,^oC$ *then* $$\left[(k_{PHA,HAc,20}^{PAO} - k_{PHA,H\,Pr,20}^{PAO}) \cdot (f_{HAc,INF}) + k_{PHA,H\,Pr,20}^{PAO} \right]$$ *where:* $$k_{PHA,HAc,20}^{PAO} = 0.80$$ $$k_{PHA,H\,Pr,20}^{PAO} = 0.33$$ $$f_{HAc,INF} = \left(S_{HAc,INF} / \left(S_{HAc,INF} + S_{H\,Prc,INF} \right) \right)$$
5	Aerobic glycogen production rate	k_{GLY}^{PAO}	*If* $T \leq 20\,^oC$ *then* $$\left[(k_{GLY,HAc,20}^{PAO} - k_{GLY,H\,Pr,20}^{PAO}) \cdot (f_{HAc,INF}) + k_{GLY,H\,Pr,20}^{PAO} \right] \cdot 1.125^{(T-20)}$$ *If* $20 < T \leq 30\,^oC$ *then* $$\left[(k_{GLY,HAc,20}^{PAO} - k_{GLY,H\,Pr,20}^{PAO}) \cdot (f_{HAc,INF}) + k_{GLY,H\,Pr,20}^{PAO} \right]$$ *where:* $$k_{GLY,HAc,20}^{PAO} = 0.015$$ $$k_{PHA,H\,Pr,20}^{PAO} = 0.077$$ $$f_{HAc,INF} = \left(S_{HAc,INF} / \left(S_{HAc,INF} + S_{H\,Prc,INF} \right) \right)$$
6	Aerobic Poly-P formation rate	k_{PP}^{PAO}	*If* $T \leq 20\,^oC$ *then* $$\left[(k_{PP,HAc}^{PAO} - k_{PP,H\,Pr}^{PAO}) \cdot (f_{HAc,INF}) + k_{PP,H\,Pr}^{PAO} \right] \cdot 1.031^{(T-20)}$$ *If* $20 < T \leq 30\,^oC$ *then* $$\left[(k_{PP,HAc}^{PAO} - k_{PP,H\,Pr}^{PAO}) \cdot (f_{HAc,INF}) + k_{PP,H\,Pr}^{PAO} \right]$$ *where:* $$k_{PHA,HAc}^{PAO} = 0.020$$ $$k_{PHA,H\,Pr}^{PAO} = 0.002$$ $$f_{HAc,INF} = \left(S_{HAc,INF} / \left(S_{HAc,INF} + S_{H\,Prc,INF} \right) \right)$$
7	Aerobic maintenance	m_{PAO}^{AER}	$$m_s^{PAO} \cdot 1.064^{(T-20)}$$

Units	Source
C-mmol/(Cmmol-h)	Oehmen et al. (2005b), Brdjanovic et al. (1998), Lopez-Vazquez et al. (2007b) This study
P-mmol/(C-mmol/h)	Smolders et al. (1994a), Brdjanovic et al. (1998), This study
C-mmol/(Cmmol-h)	Brdjanovic et al. (1998), This study
C-mmol/(Cmmol-h)	Meijer et al. (2002), This study
P-mmol/(Cmmol-h)	Brdjanovic et al. (1997), Brdjanovic et al. (1998), This study
C-mmol/(C-mmol/h)	Brdjanovic et al. (1997)

Kinetic parameters for *Competibacter*.

	Process		Parameter

8	Anaerobic acetate uptake rate	$q_{Competi,HAc}^{MAX}$	*If T ≤ 20 °C then*
			
			$q_{Competi,HAc,20}^{MAX} \cdot 1.211^{(T-20)} \cdot \dfrac{10^{-pH}}{1.2x10^{-8}+10^{-pH}}$
			
			If 20 < T ≤ 30 °C then
			
			$0.24 \cdot \dfrac{10^{-pH}}{1.2x10^{-8}+10^{-pH}}$
			
			where:
			
			$q_{Competi,HAc,20}^{MAX} = 0.22$

9	Anaerobic propionate uptake rate	$q_{Competi,H\,Pr}^{MAX}$	*If T ≤ 20 °C then*
			
			$q_{Competi,H\,Pr,20}^{MAX} \cdot 1.211^{(T-20)} \cdot \dfrac{10^{-pH}}{1.2x10^{-8}+10^{-pH}}$
			
			If 20 < T ≤ 30 °C then
			
			$q_{Competi,H\,Pr,20}^{MAX} \cdot \dfrac{10^{-pH}}{1.2x10^{-8}+10^{-pH}}$
			
			where:
			
			$q_{Competi,H\,Pr,20}^{MAX} = 0.01$

10	Anaerobic maintenance rate	$m_{Competi}^{ANA}$	$0.5 \cdot m_{ATP,Competi}^{ANA} \cdot 1.028^{(T-20)}$ *where:* $m_{ATP,PAO,20}^{ANA} = 2.35x10^{-3}\,ATP-mol/(C-mol-h)$
11	Aerobic PHA degradation	$k_{PHA}^{Competi}$	*If 10 ≤ T ≤ 30 °C then* $0.80 \cdot k_{PHA,20}^{Competi} \cdot 1.109^{(T-20)}$ *where:* $k_{PHA,20}^{Competi} = 1.10$
12	Aerobic glycogen production	$k_{GLY}^{Competi}$	*If 10 ≤ T ≤ 30 °C then* $k_{GLY,20}^{Competi} \cdot 1.071^{(T-20)}$ *where:* $k_{GLY,20}^{Competi} = 0.30$
13	Aerobic maintenance	$m_{Competi}^{AER}$	$m_{S}^{Competi} \cdot 1.046^{(T-20)}$

Units	Source
C-mmol/(Cmmol-h)	Filipe et al. (2001a) Oehmen et al. (2005b), Lopez-Vazquez et al. (2007b) This study
C-mmol/(Cmmol-h)	Filipe et al. (2001a) Oehmen et al. (2005b), Lopez-Vazquez et al. (2007b) This study
C-mmol/(Cmmol-h)	Smolders et al. (1994b) Filipe et al. (2001a) Zeng et al. (2003a) Lopez-Vazquez et al. (2007b)
C-mmol/(Cmmol-h)	Lopez-Vazquez et al. (2008b,c) This study
C-mmol/(Cmmol-h)	Lopez-Vazquez et al. (2008b) This study
C-mmol/(C-mmol/h)	Lopez-Vazquez et al. (2008b)

Kinetic parameters for *Alphaproteobacteria-GAO.*

	Process		Parameter

14	Anaerobic acetate uptake rate	$q_{Alpha,HAc}^{MAX}$	*If $10 \leq T \leq 20\,^{o}C$ then*
			
			$q_{Alpha,HAc,20}^{MAX} \cdot 1.211^{(T-20)} \cdot \dfrac{10^{-pH}}{1.2x10^{-8}+10^{-pH}}$
			
			If $20 < T \leq 30\,^{o}C$ then
			
			$0.13 \cdot \dfrac{10^{-pH}}{1.2x10^{-8}+10^{-pH}}$
			
			where:
			$q_{Alpha,HAc,20}^{MAX}=0.11$

15	Anaerobic propionate uptake rate	$q_{Alpha,H\,Pr}^{MAX}$	*If $10 \leq T \leq 20\,^{o}C$ then*
			
			$q_{Alpha,H\,Pr,20}^{MAX} \cdot 1.211^{(T-20)} \cdot \dfrac{10^{-pH}}{1.2x10^{-8}+10^{-pH}}$
			
			If $20 < T \leq 30\,^{o}C$ then
			
			$0.24 \cdot \dfrac{10^{-pH}}{1.2x10^{-8}+10^{-pH}}$
			
			where:
			$q_{Alpha,H\,Pr,20}^{MAX}=0.22$

16	Anaerobic maintenance rate	m_{Alpha}^{ANA}	$0.5 \cdot m_{ATP,Alpha}^{ANA} \cdot 1.028^{(T-20)}$ *where:* $m_{ATP,PAO,20}^{ANA}=2.35x10^{-3}\,ATP-mol\,/(C-mol-h)$
17	Aerobic PHA degradation	k_{PHA}^{Alpha}	*If $10 \leq T \leq 30\,^{o}C$ then* $0.80 \cdot k_{PHA,20}^{Alpha} \cdot 1.109^{(T-20)}$ *where:* $k_{PHA,20}^{Alpha}=0.50$
18	Aerobic glycogen production	k_{GLY}^{Alpha}	*If $10 \leq T \leq 30\,^{o}C$ then* $k_{GLY,20}^{Alpha} \cdot 1.071^{(T-20)}$ *where:* $k_{GLY,20}^{Alpha}=0.06$
19	Aerobic maintenance	m_{Alpha}^{AER}	$m_{S}^{Alpha} \cdot 1.046^{(T-20)}$

Units	Source
C-mmol/(Cmmol-h)	Filipe et al. (2001a) Oehmen et al. (2005a), Oehmen et al. (2006b), Lopez-Vazquez et al. (2007b) This study
C-mmol/(Cmmol-h)	Filipe et al. (2001a) Oehmen et al. (2005a), Oehmen et al. (2006b), Lopez-Vazquez et al. (2007b) This study
C-mmol/(Cmmol-h)	Smolders et al. (1995) Filipe et al. (2001a) Zeng et al. (2003a) Lopez-Vazquez et al. (2007b)
C-mmol/(Cmmol-h)	Lopez-Vazquez et al. (2008b,c) This study
	Lopez-Vazquez et al. (2008b) This study
C-mmol/(C-mmol/h)	Lopez-Vazquez et al. (2008b)

Appendix 7.6

Kinetic expressions for *Accumulibacter*.

Process	Expression
1 Anaerobic acetate uptake	$q_{PAO,VFA}^{MAX} \cdot \dfrac{S_{HAc}}{(S_{HAc} + K_{S,HAc})} \cdot X_{PAO}$
2 Anaerobic propionate uptake	$q_{PAO,VFA}^{MAX} \cdot \dfrac{S_{HPr}}{(S_{HPr} + K_{S,HPr})} \cdot X_{PAO}$
3 Anaerobic maintenance	$m_{PAO}^{ANA} \cdot X_{PAO}$
4 Aerobic PHA degradation	$k_{PHA}^{PAO} \cdot f_{PAO,PHA}^{2/3} \cdot X_{PAO}$
5 Aerobic glycogen production	$k_{GLY}^{PAO} \cdot f_{PAO,PHA}^{2/3} \cdot \left(\dfrac{1}{f_{PAO,GLY}} \right) \cdot X_{PAO}$
6 Aerobic Poly-P formation	$k_{PP}^{PAO} \cdot \left(\dfrac{1}{f_{PAO,PP}} \right) \cdot \dfrac{\left(f_{PAO,PP}^{MAX} - f_{PAO,PP} \right)}{\left(f_{PAO,PP}^{MAX} - f_{PAO,PP} \right) + K_{S,GLY}} \cdot X_{PAO}$
7 Aerobic maintenance	$m_{PAO}^{AER} \cdot X_{PAO}$

Switch function	Source
$$\frac{X_{PP}}{X_{PP}+K_{S,PP}}\cdot\frac{X_{GLY}}{X_{GLY}+K_{S,GLY}}\cdot\left[1-\frac{S_{O2}}{S_{O2}+K_{S,O2}}\right]$$	Murnleitner et al. (1997), Meijer et al. (2002)
$$\frac{X_{PP}}{X_{PP}+K_{S,PP}}\cdot\frac{X_{GLY}}{X_{GLY}+K_{S,GLY}}\cdot\left[1-\frac{S_{O2}}{S_{O2}+K_{S,O2}}\right]$$	Murnleitner et al. (1997), Meijer et al. (2002)
$$\frac{X_{PP}}{X_{PP}+K_{S,PP}}\cdot\left[1-\frac{S_{O2}}{S_{O2}+K_{S,O2}}\right]$$	Murnleitner et al. (1997)
$$\frac{f_{PAO,PHA}}{f_{PAO,PHA}+K_{S,fPHA}}\cdot\frac{S_{SPO4}}{S_{SPO4}+K_{S,SPO4}}\cdot\frac{S_{O2}}{S_{O2}+K_{S,O2}}$$	Murnleitner et al. (1997), Meijer et al. (2002)
$$\frac{\left(f_{PAO,GLY}^{MAX}-f_{PAO,GLY}\right)}{\left(f_{PAO,GLY}^{MAX}-f_{PAO,GLY}\right)+K_{S,GLY}}\cdot\frac{X_{PHA}^{PAO}}{X_{PHA}^{PAO}+K_{S,PHA}}\cdot\frac{S_{O2}}{S_{O2}+K_{S,O2}}$$	Murnleitner et al. (1997)
$$\frac{X_{PHA}^{PAO}}{X_{PHA}^{PAO}+K_{S,PHA}}\cdot\frac{S_{SPO4}}{S_{SPO4}+K_{S,SPO4}}\cdot\frac{S_{O2}}{S_{O2}+K_{S,O2}}$$	Murnleitner et al. (1997), Meijer et al. (2002)
$$\frac{S_{O2}}{S_{O2}+K_{S,O2}}$$	Murnleitner et al. (1997)

Kinetic expressions for *Competibacter.*

	Process	Expression
8	Anaerobic acetate uptake	$q_{Competi,HAc}^{MAX} \cdot \dfrac{S_{HAc}}{(S_{HAc} + K_{S,HAc})} \cdot X_{Competi}$
9	Anaerobic propionate uptake	$q_{Competi,HPr}^{MAX} \cdot \dfrac{S_{HPr}}{(S_{HPr} + K_{S,HPr})} \cdot X_{Competi}$
10	Anaerobic maintenance	$m_{Competi}^{ANA} \cdot X_{Competi}$
11	Aerobic PHA degradation	$k_{PHA}^{Competi} \cdot f_{Competi,PHA}^{2/3} \cdot X_{Competi}$
12	Aerobic glycogen production	$k_{GLY}^{Competi} \cdot f_{Competi,PHA}^{2/3} \cdot \left(\dfrac{1}{f_{Competi,GLY}} \right) \cdot X_{Competi}$
13	Aerobic maintenance	$m_{Competi}^{AER} \cdot X_{Competi}$

Switch function	Source
$$\frac{X_{GLY}^{Competi}}{X_{GLY}^{Competi} + K_{S,GLY}} \cdot \left[1 - \frac{S_{O2}}{S_{O2} + K_{S,O2}} \right]$$	Murnleitner et al. (1997), Meijer et al. (2002)
$$\frac{X_{GLY}^{Competi}}{X_{GLY}^{Competi} + K_{S,GLY}} \cdot \left[1 - \frac{S_{O2}}{S_{O2} + K_{S,O2}} \right]$$	Murnleitner et al. (1997), Meijer et al. (2002)
$$\frac{X_{GLY}^{Competi}}{X_{GLY}^{Competi} + K_{S,GLY}} \cdot \left[1 - \frac{S_{O2}}{S_{O2} + K_{S,O2}} \right]$$	Murnleitner et al. (1997)
$$\frac{f_{Competi,PHA}}{f_{Competi,PHA} + K_{S,fPHA}} \cdot \frac{S_{SPO4}}{S_{SPO4} + K_{S,SPO4}} \cdot \frac{S_{O2}}{S_{O2} + K_{S,O2}}$$	Murnleitner et al. (1997), Meijer et al. (2002), This study
$$\frac{\left(f_{Competi,GLY}^{MAX} - f_{Competi,GLY} \right)}{\left(f_{Competi,GLY}^{MAX} - f_{Competi,GLY} \right) + K_{S,GLY}} \cdot \frac{X_{PHA}^{Competi}}{X_{PHA}^{Competi} + K_{S,PHA}} \cdot \frac{S_{O2}}{S_{O2} + K_{S,O2}}$$	Murnleitner et al. (1997), This study
$$\frac{S_{O2}}{S_{O2} + K_{S,O2}}$$	Murnleitner et al. (1997), This study

Kinetic expressions for *Alphaproteobacteria-GAO.*

	Process	Expression
14	Anaerobic acetate uptake	$q_{Alpha,HAc}^{MAX} \cdot \dfrac{S_{HAc}}{(S_{HAc} + K_{S,HAc})} \cdot X_{Alpha}$
15	Anaerobic propionate uptake	$q_{Alpha,HPr}^{MAX} \cdot \dfrac{S_{HPr}}{(S_{HPr} + K_{S,HPr})} \cdot X_{Alpha}$
16	Anaerobic maintenance	$m_{Alpha}^{ANA} \cdot X_{Alpha}$
17	Aerobic PHA degradation	$k_{PHA}^{Alpha} \cdot f_{Alpha,PHA}^{2/3} \cdot X_{Alpha}$
18	Aerobic glycogen production	$k_{GLY}^{Alpha} \cdot f_{Alpha,PHA}^{2/3} \cdot \left(\dfrac{1}{f_{Alpha,GLY}} \right) \cdot X_{Alpha}$
19	Aerobic maintenance	$m_{Alpha}^{AER} \cdot X_{Alpha}$

Switch function	Source
$\dfrac{X_{GLY}^{Alpha}}{X_{GLY}^{Alpha} + K_{S,GLY}} \cdot \left[1 - \dfrac{S_{O2}}{S_{O2} + K_{S,O2}} \right]$	Murnleitner et al. (1997), Meijer et al. (2002)
$\dfrac{X_{GLY}^{Alpha}}{X_{GLY}^{Alpha} + K_{S,GLY}} \cdot \left[1 - \dfrac{S_{O2}}{S_{O2} + K_{S,O2}} \right]$	Murnleitner et al. (1997), Meijer et al. (2002)
$\dfrac{X_{GLY}^{Alpha}}{X_{GLY}^{Alpha} + K_{S,GLY}} \cdot \left[1 - \dfrac{S_{O2}}{S_{O2} + K_{S,O2}} \right]$	Murnleitner et al. (1997)
$\dfrac{f_{Alpha,PHA}}{f_{Alpha,PHA} + K_{S,fPHA}} \cdot \dfrac{S_{SPO4}}{S_{SPO4} + K_{S,SPO4}} \cdot \dfrac{S_{O2}}{S_{O2} + K_{S,O2}}$	Murnleitner et al. (1997), Meijer et al. (2002), This study
$\dfrac{\left(f_{Alpha,GLY}^{MAX} - f_{Alpha,GLY} \right)}{\left(f_{Alpha,GLY}^{MAX} - f_{Alpha,GLY} \right) + K_{S,GLY}} \cdot \dfrac{X_{PHA}^{Alpha}}{X_{PHA}^{Alpha} + K_{S,PHA}} \cdot \dfrac{S_{O2}}{S_{O2} + K_{S,O2}}$	Murnleitner et al. (1997), This study
$\dfrac{S_{O2}}{S_{O2} + K_{S,O2}}$	Murnleitner et al. (1997), This study

Appendix 7.7.

Kinetic coefficients for biomass.

Kinetic coefficient	Description	Value
$K_{S,HAc}$	Half-saturation coefficient for acetate	0.001
$K_{S,HPr}$	Half-saturation coefficient for propionate	0.001
$K_{S,PHA}$	Half-saturation coefficient for PHA	0.01
$K_{S,GLY}$	Half-saturation coefficient for glycogen	0.01
$K_{S,PP}$	Half-saturation coefficient for poly-phosphate (poly-P)	0.01
$K_{S,PO4}$	Half-saturation coefficient for orthophosphate	0.01
$K_{S,O2}$	Half-saturation coefficient for oxygen	0.01
$K_{S,fPHA}$	Half-saturation coefficient for the fraction of PHA in biomass	0.01
$f_{Xi,Pi} = \dfrac{X_{Pi}^{Xi}}{X_i}$	Intracellular fraction of compound P_i (PHA, glycogen or poly-P) in biomass X_i (*Accumulibacter, Competibacter* or *Alphaproteobacteria*)	
$f_{PAO,PP}^{MAX}$	Maximum poly-P fraction in *Accumulibacter* biomass	0.30
$f_{PAO,GLY}^{MAX}$	Maximum glycogen fraction in *Accumulibacter* biomass	0.27
$f_{Competi,GLY}^{MAX}$	Maximum glycogen fraction in *Competibacter* biomass	0.35
$f_{Alpha,GLY}^{MAX}$	Maximum glycogen fraction in *Alphaproteobacteria* biomass	0.35

Units	Source
C-mmol/L	Switch
C-mmol/L	Switch
C-mmol/L	Switch
C-mmol/L	Switch
P-mmol/L	Switch
P-mmol/L	Switch
O_2-mmol/L	Switch
C-mmol/C-mmol	Switch
P-mmol/C-mmol	Wentzel et al. (1989)
C-mmol/C-mmol	Smolders et al. (1995)
C-mmol/C-mmol	Lopez-Vazquez et al. (2008)
C-mmol/C-mmol	Lopez-Vazquez et al. (2008)

8
Chapter

General conclusions, evaluation and outlook

Content

8.1. General conclusions

Aiming at limiting the phosphorus loads released into the environment through the treated sewerage, there is an increasing need to keep well-controlled, operated and efficient phosphorus removal processes in wastewater treatment systems. Being enhanced biological phosphorus removal (EBPR) in activated sludge wastewater treatment plants (WWTP) the preferred and most widely applied phosphorus removal technology, it becomes indispensable to understand how the interaction among the different involved microbial communities affects the process performance and stability.

Despite that big efforts have been made from different angles and perspectives embracing scientists, engineers and plant practitioners, a better understanding about the environmental and operating factors affecting the competition between polyphosphate-accumulating organisms (PAO) and glycogen-accumulating organisms (GAO) is needed for the development of control measures that may ultimately lead to the operation of more stable and reliable EBPR systems.

Temperature, carbon source and pH have been pointed out as important parameters to understand the dominance and appearance of PAO and GAO under different environmental and operating conditions. However, important aspects like the actual temperature influence on the metabolism of GAO (on the anaerobic and aerobic stoichiometry and kinetics) or about the combined effect of key factors were not known. Furthermore, excluding a few full-scale studies, most of the research has mainly been carried out at lab-scale, without looking into full-scale EBPR WWTP in order to find out more about the conditions influencing the PAO and GAO microbial communities.

Through undertaking different studies at both lab- and full-scale and by applying mathematical modeling, in the PhD research presented in this dissertation, the factors influencing the PAO-GAO competition have been addressed from different perspectives.

The short- (hours) and long-term (weeks) effects of temperature on the metabolism of GAO regarding the stoichiometry and kinetics of the different anaerobic and aerobic microbial conversions were evaluated. For this purpose, an enriched GAO culture was cultivated in a lab-scale sequencing batch reactor (SBR) and anaerobic and aerobic batch tests were executed within a relatively wide temperature range (from 10 to 40 °C) that covers the operating temperature of most of the municipal and industrial WWTP.

In order to evaluate the environmental and operating conditions that influence the occurrence of PAO and GAO at full-scale EBPR WWTP, a full-scale survey in seven treatment plants in The Netherlands was undertaken under cold weather conditions (average sewage temperature around 12 °C). Operating data were collected, lab-batch tests were executed to determine the EBPR biomass activity and the PAO and GAO microbial communities were quantified using Fluorescence *in situ* Hybribization (FISH). Through a statistical analysis, significant correlations were looked for among diverse operating and environmental conditions and the occurrence and activity of PAO and GAO populations.

Using lab-enriched cultures, a practical method for the quantification of PAO and GAO in full-scale EBPR WWTP based on commonly performed analytical determinations (such as acetate and orthophosphate) was developed and validated on samples from full-scale systems.

By compiling the metabolic models of PAO (*Accumulibacter*) and GAO (as *Competibacter* and *Alpha-proteobacteria*-GAO) as well as their carbon, temperature and pH-dependences reported in literature in a mechanistic mathematical model, the combined effects of type of carbon source, temperature and pH on the PAO-GAO competition were studied.

The main conclusions from the present study are:

1.   Regarding the short-term temperature effects on GAO

Based on the anaerobic and aerobic temperature dependences of their metabolisms (stoichiometry and kinetics), GAO are only able to compete with PAO for substrate at temperatures higher than 20 °C (at pH 7.0 and 8 d SRT). Below this temperature, GAO do not have metabolic advantages over PAO. The temperature dependences of GAO showed to have a moderate or medium degree of dependence on temperature.

2.   On the long-term temperature effects on GAO

Although the long-term substrate uptake rate of GAO above 20 °C was lower than in short-term studies, it was still higher than that of PAO (around 12 %) showing that GAO have kinetic advantages and may tend to proliferate at temperatures higher than 20 °C. At temperatures lower than 20 °C (e.g. 10 °C), the anaerobic metabolism of GAO, in particular the anaerobic glycogen hydrolysis, resulted inhibited limiting the substrate uptake rate and, therefore, the growth of GAO. A complete switch in the dominant microbial populations from an enriched GAO to an enriched PAO culture at 10 °C confirmed that GAO cannot adapt to low temperatures being unable to compete with PAO at lower temperatures.

The net biomass growth rate of GAO, which can be assumed to be a resultant of the overall temperature effects on their anaerobic and aerobic metabolism, was highly sensitive to temperature changes. Interestingly and as previously observed with PAO, under aerobic conditions GAO tend to favor the storage of intracellular polymers, such as glycogen, over growth. A calculation of the minimum aerobic solids retention time (SRT) of GAO showed that these organisms have a lower biomass growth rate than PAO requiring a longer SRT from 10 to 30 °C, particularly below 20 °C. On the other hand, due to strong increases in the aerobic maintenance requirements and less efficient energy production (oxidative phosphorylation activity), the net aerobic growth of GAO was limited at high temperatures (30 and 40 °C), requiring also longer SRT above 30 °C than that at 20 °C.

3.   Concerning the occurrence of PAO and GAO at full-scale systems
The occurrence of PAO (*Accumulibacter*) was positively correlated with pH values, suggesting that high pH levels are favorable for these microorganisms at full-scale systems. The appearance of PAO was also influenced by the operation of well-defined denitrification stages that, furthermore, also stimulated the development of denitrifying PAO (DPAO). Quantified GAO populations (*Competibacter*) were only correlated with the organic matter concentrations implying that, according to the conditions of this survey, these microorganisms are only able to grow on the excess of organic matter (or substrate) in the system. *Defluviccocus*-related microorganisms and *Sphingomonas* were not observed or only seen in negligible fractions in a few plants (< 1%) indicating that these organisms could be rarely present in municipal EBPR WWTP and, thus, may not be the main competitors of PAO.

4.   About the practical method for the quantification of PAO and GAO
Through executing an anaerobic batch test and measuring commonly determined parameters such as acetate, orthophosphate and mixed liquor suspended solids, a practical method showed to be potentially suitable for the *in situ* quantification of PAO and GAO populations at full-scale systems.

5.   On modeling the PAO-GAO competition
At low temperature (10 °C), PAO were the dominant microorganisms since the metabolism of GAO was inhibited at 10 °C. At moderate temperature (20 °C), the simultaneous presence of acetate and propionate as carbon sources (e.g. 75-25 or 50-50 % acetate to propionate ratios) favored the growth of PAO over GAO, regardless of the applied pH. In contrast, when either acetate or propionate is supplied as sole carbon source at 20 °C, PAO is only favored over GAO when a high pH (7.5) is applied. In general, GAO (*Competibacter* and *Alphaproteobacteria-GAO*) tended to proliferate at higher temperature (30 °C). Nevertheless, an adequate acetate to propionate

ratio (75 – 25 %) and a pH not lower than 7.0 could be used as potential tools to suppress the proliferation of GAO at high temperature (30 °C).

8.2. Evaluation and outlook

8.2.1. On the temperature effects on the PAO-GAO competition

On the basis of this study, GAO can only compete with PAO at temperatures higher than 20 °C. Below 20 °C, the anaerobic glycogen conversion appeared to limit the proliferation of GAO (e.g. 10 °C). Therefore, operation of EBPR lab-scale reactors at a low temperature, such as 10 °C, could be used as a strategy to suppress the growth of GAO aiming at obtaining highly enriched PAO cultures. This strategy could also be possibly applied when looking after the isolation of PAO. On the other hand, the proliferation of GAO in EBPR systems could be suppressed through adjusting (shortening) the applied SRT to the minimum aerobic SRT of PAO as a function of the applied temperature. Thus, besides operating the lab-reactors at 10 °C, more favourable conditions for PAO could also be created through shortening the SRT.

In the present study, the temperature effects on GAO were calculated from 10 to 40 °C. However, a systematic study on an enriched PAO culture at temperatures higher than 30 °C has not been performed. Such a study could be very useful to address the feasibility of achieving EBPR at higher temperatures like in warm climate countries or in WWTP treating industrial sewerage. In certain reports the operation of EBPR systems above 30 °C and even up to 37 °C has been evaluated (Jones *et al.*, 1987; Yeoman *et al.*, 1988; McClinton *et al.*, 1993; Converti *et al.*, 1995; Mamais and Jenkins, 1992). However, the main conclusions from those reports tend to be inconsistent. Likely, the growth and proliferation of GAO in some of those studies led to deviated observations and discrepancies. Thus, the main challenge might be to suppress the growth of GAO at this temperature range. Nevertheless, at temperatures higher than 30 °C, GAO require longer minimum aerobic SRT. If the minimum SRT of GAO is higher than that of PAO, there may be possibilities for the implementation and operation of EBPR process at WWTP operated at higher temperatures. Long-term lab- and full-scale tests will be useful to address this potential strategy.

Despite that *Competibacter* and *Alphaproteobacteria*-GAO (e.g. *Defluviicoccus vanus* related microorganisms) share a similar physiology, further research is needed to address the temperature dependencies of the different conversion rates involved in the metabolism of *Alphaproteobacteria-GAO*. Moreover, to perform similar temperature-effect studies on an enriched *Accumulibacter* culture grown on propionate is also recommendable. This will be relevant to evaluate and compare the temperature dependencies of PAO cultures grown on acetate and propionate.

Interestingly, the anaerobic and aerobic kinetics of GAO showed to have different temperature dependencies at short- and long-term as a consequence of the overall long-term temperature effects (particularly regarding the maximum substrate uptake rate). These effects are highly sensitive since a single change on any of the metabolic conversions tends to be magnified and reflected on the overall biomass activity due to the close relationship between the anaerobic and aerobic metabolisms as the end of one phase (either anaerobic or aerobic) defines the start of the other (because of the alternating anaerobic-aerobic conditions). Therefore, whenever possible, long-term studies are desirable to address potential biomass acclimations, adaptations or changes. Nevertheless, short-term studies seem to be feasible when studying temporary changes caused by fluctuations on the operating and environmental conditions or when evaluating the potential exposure of biomass for short periods of time to particular circumstances (e.g. sudden discharges of toxic compounds).

8.2.2. On full-scale studies

While most of the research has focused on lab-scale research, little is still known about which and how different factors affect the occurrence and appearance of PAO and GAO populations in full-scale EBPR WTTP. In the present research, a survey in different treatment plants in The Netherlands was carried out, finding important correlations to understand the stability and reliability of these systems. However, a similar study undertaken under summer conditions with average sewerage temperature higher than 20 °C would be rather useful to address the effects of higher temperatures on the PAO and GAO microbial populations. Furthermore, EBPR WWTP receiving or treating industrial discharges on a regular basis (comprising more than 20 − 25 % of the total WWTP influent) could also be included not only because of the higher temperature but also because industrial wastewater is usually more complex containing diverse types of (slowly and easily biodegradable) organic matter. This may have a strong influence on the occurrence, appearance and diversity of different PAO and GAO strains (e.g. *Actino*-PAO, *Defluviicoccus*-related microorganisms and *Sphingomonas*). The use of advanced molecular techniques, like FISH-Microautoradiography, could help to define the active microbial populations. These results together with historical data of operating and environmental conditions from the treatment plants and biomass activity data obtained from lab-scale batch tests could bring important and useful information to study the factors influencing the PAO-GAO competition at full-scale EBPR WWTP.

8.2.3. On the quantification of PAO and GAO

From an operational perspective, a fast, simple, economical and reliable method to quantify PAO and GAO could be an important and useful tool to monitor the day-a-day operating conditions at EBPR plants. In the present study, a practical method for the quantification of PAO and GAO was proposed. It proved suitable for the quantification of these microbial populations at full-scale systems. Nevertheless, the method requires to be tested in more treatment plants and under different environmental and operating conditions to assess its reliability. Moreover, following a similar methodology, it could be extended using highly enriched cultures of PAO and GAO grown on other carbon sources (e.g. propionate, ethanol, methanol, glucose, aminoacids). Compared to advanced molecular techniques (like FISH and FISH-MAR), the benefit of developing such practical methods is that, in a fast, simple and economic way, only active biomass responding to the added substrate will be quantified increasing its precision and reliability.

8.2.4. On modeling the PAO-GAO competition

Metabolic modeling can be a highly useful means of describing key factors affecting the complex interactions between different microorganisms relevant in EBPR systems, and in predicting their resulting population dynamics. Incorporating the main factors affecting the PAO-GAO competition that are described in this study into future modeling endeavors focused on the optimization of full-scale EBPR plants would likely facilitate improved process efficiency and robustness.

Since the PAO-GAO competition is highly dependent upon the carbon source, and in particular on the influent acetate to propionate ratio, future modeling efforts could be also directed towards modeling the anaerobic fermentation processes taking place on pre-fermentors and in the anaerobic stages of EBPR WWTP. If an anaerobic model could be integrated together with the PAO and GAO models, it would likely help to optimize the operation of EBPR plants.

8.2.5. On the factors influencing the PAO-GAO competition

Using mathematical modelling, the combined effects of temperature, carbon source and pH were evaluated. On the basis of the present study, to supply an adequate acetate to propionate ratio (75 - 25 %) and a pH not lower than 7.0 appears to be sufficient for the enrichment of *Accumulibacter,* limiting the development of GAO (either *Competibacter* or *Alphaproteobacteria*-GAO) regardless of the sewage temperature. Even at 30 °C, these conditions appear to be sufficient to suppress the growth of GAO. From a practical and operational perspective, to provide an influent 75 to 25 % acetate to propionate ratio at full-scale systems and ensure a pH level higher than 7.0

seems to be feasible (e.g. through modifying the operating conditions of the pre-fermentors and by adjusting the pH through acid and based addition in a tank equalizer). Future research is needed regarding the operation of pre-fermenters to provide certain influent volatile fatty acids fractions and their influence on the PAO-GAO competition at full-scale EBPR systems.

8.2.6. On other EBPR related issues

In different studies (e.g. Schuler and Jenkins, 2003), casaminoacids were supplied as an additional carbon source. The main rationale behind its use was the apparent correlation between the supply of casaminoacids and the stable operation of lab-scale EBPR systems. When casaminoacids were not fed in the influent, certain lab-scale systems lost their EBPR activity, which was recovered when they were supplied again (Schuler, personal communication). So far, not too much is known about the effects of this complex carbon source on the metabolism of PAO. Moreover, a wide range of carbon sources (likely including casaminoacids) can be expected in the influent of WWTP. Could casaminoacids or hydrolyzed by-products of them be only metabolized by PAO and not by GAO? Kong et al. (2005) found that certain strains of *Actinobacteria* are able to store a mixture of aminoacids under anaerobic conditions and remove phosphorus aerobically. While *Accumulibacter* has been strongly pointed out as a PAO, it appears that they may not be the only PAO present in EBPR systems.

It is important to notice that most of the lab-scale EBPR research is carried out at around 20 °C and neutral pH (7.0) which appears to be close to the breaking points or borderlines between the dominance of PAO and GAO. Thus, the sensitivity and variability of pH probes, water baths and thermostats could unintentionally tend to create slightly favourable conditions for either PAO or GAO at long-term, which in the latter case may lead to the unexpected proliferation of GAO.

Phosphorus is a finite natural resource, which plays an important role in nature. In most of the municipal and industrial wastewater treatment plants, it is removed from the water phase and concentrated in the sludge in secondary clarifiers. Due to stricter legislation that prohibit the use of dewatered sludge for agricultural purposes, this sludge is dewatered and disposed of through incineration. Thus, phosphorus goes out of the natural environment and perhaps gets confined in the structure of civil works or landfills. The implementation and operation of wastewater treatment configurations which open the possibility of phosphorus recovery, such as the BCFS and the Phostrip process (van Loosdrecht et al., 1998; Brdjanovic et al., 2000), should be carefully analyzed in order to ensure that the future of phosphorus in the world would not be jeopardized by the current treatment technologies.

References

Brdjanovic D, Logemann S, van Loosdrecht MCM, Hooijmans CM, Alaerts GJ, Heijnen JJ (1998) Influence of temperature on biological phosphorus removal: process and molecular ecological studies. Water Res 32(4):1035-1048.

Brdjanovic D, van Loosdrecht MCM, Hooijmans CM, Alaerts GJ, Heijnen JJ (1997) Temperature effects on physiology of biological phosphorus removal. ASCE J Environ Eng 123(2):144-154.

Brdjanovic D, van Loosdrecht MCM, Versteeg P, Hooijmans CM, Alaerts GJ, Heijnen JJ (2000) Modeling COD, N and P removal in a full-scale wwtp Haarlem Waarderpolder. Water Res. 34, 846–858.

Converti A, Robatti M, Broghi del M (1995) Biological removal of phosphorus from wastewaters by alternating aerobic and anaerobic conditions. Wat Res 29(1):263-269.

Jones PH, Tadwalkar AD, Hsu CL (1987) Enhanced uptake of phosphorus by activated sludge – effect of substrate addition. Wat Res 21(5):301-308.

Kong Y, Nielsen JL, Nielsen PH (2005) Identity and ecophysiology of uncultured actinobacterial polyphosphate-accumulating organisms in full-scale enhanced biological phosphorus removal plants. Appl. Environ. Microbiol. 71, 4076 - 4085.

Mamais D, Jenkins D (1992) The effects of MCRT and temperature on enhanced biological phosphorus removal. Wat Sci Technol 26(5-6): 955-965.

McClinton SA, Randall CW, Patterkine VM (1993) Effects of temperature and mean cell residence time on biological nutrient removal processes. Wat Environ Res 65(5):110-118.

Schuler AJ, Jenkins D (2003) Enhanced biological phosphorus removal from wastewater by biomass with different phosphorus contents, Part 1: Experimental results and comparison with metabolic models. Water Environ. Res. 75, 485–498.

van Loosdrecht MCM, Brandse FA, Vries AC (1998) Upgrading of wastewater treatment processes for integrated nutrient removal – The BCFS process. Water Sci. Technol. 37, 209–217.

Yeoman S, Hunter M, Stephenson T, Lester JN, Perry R (1988) An assessment of excess biological phosphorus removal during activated sludge treatment. Environ Technol Letter 9:637-646.

Acknowledgements

This seems to be the end of a marvelous journey started in 2004, and perhaps several months before... Since then, many people have got involved not only in the research that led to the dissertation that now you hold on your hands but also in my life. Unforgettable moments and experiences from Delft, one of the most *gellezig* places on Earth.

I would like to acknowledge my sponsor, the Mexican National Council for Science and Technology (CONACYT), for the grant awarded to carry out my PhD research. Special thanks to my *alma mater* the Autonomous University of the State of Mexico for all the support and motivation to give the first steps (in that moment the most difficult ones) to start this journey.

Difficult to deny how lucky and fortunate I have been to work and been supervised by a great and remarkable group of people. First, my most sincere thanks to my promoters Prof. Mark van Loosdrecht and Prof. Huub Gijzen. Mark, you have been always inspiring, motivating, and an endless source of ideas... thank you very much for all your patience and guidance but foremost for being always there to give an advice about Bio-P (one of your greatest passions) or at personal level. Huub, you opened me the door to come to Delft (I can not forget those first e-mails that we exchanged) and later on struggled to set and 'shape' the foundations of this research. Unfortunately, you had to leave in the middle of this trip but managed to be in contact and keep and eye on my (our) research with your always critical and constructive comments. Thanks a lot! Also, I am very grateful to my supervisors Dr. Tineke Hooijmans and Dr. Damir Brdjanovic. Tineke, it is difficult to find the words that could best describe all what you have done for me, your endless help, support and priceless friendship. You were always there to listen to me and stood behind me when I needed. Thank you very much! Damir, I appreciate to have worked with you, your ideas, (entrepreneurial) vision and proactivity which had a big impact on the main outcome of this research. Thanks a lot! Thanks also to Dr. Moustafa Moussa for helping me to settle down (academically and experimentally) at UNESCO-IHE during my first months in Delft.

I would like to thank the staff of the Environmental Resources Department, in particular UNESCO-IHE lab staff for all your continuous support and assistance (many thanks to Fred, Frank, Lyz, Don and Peter!) as well as Vera, Henk and Diederick, for sharing so many smiles, jokes and points of views about so many (philosophical, cultural, sport and historical) different topics mostly during those unforgettable coffee breaks. Thanks also to Jolanda Boots from Student Affairs department for always finding the way to solve the most unpredictable (financial) situation. My most sincere appreciation to Young Il-Song, Joost van den Bulk and Jos van der Ent, MSc and BSc students, who kindly collaborated in this research. I know you all are doing fine. Wish you all the best in your careers!

Thanks a lot to Patricia, Margarida, Merle, Marlies, Udo, Sjaak, Max, Mario, Katja, Yang, Cor, but specially to Geert, from the Department of Biotechnology at TUDelft for all your help to introducing me into the marvelous biotech world, providing me analytical protocols, assisting me to analyze my (hundreds) samples, as well as helping me to get and set-up diverse lab equipment.

A sincere word of appreciation to Dr. Adrian Oehmen and Prof. Zhiguo Yuan for all their openness and professionalism that led to a fruitful collaboration and a new friendship.

After all these years in Delft, thanks a lot to all my friends coming from everywhere (Latin-America, Asia, Africa and Europe) with whom we have shared laughs, tears, parties, dinners and trips that will remain forever in my mind making my stay in The Lowlands a very pleasant one. Thanks to Eduardo and Nicole, Humberto, Pato and Mae, Mara, Rolando, Kro, Margarita, Andrés (Delft was not the same without you, guys!), Victor, Alejandro and Xavier (and those unforgettable beer trips and our (clandestine) beer factory...!), Nelson, Felipe (Jelipillo, qué paso huerco?!), Ali, Assiyeh, Shilp, Jaime, Sylvie, Javier and Zaira, Leo, Sandra and Vale, Sarita and Thomas, Isra, Oswa, Shana and Camila, Ioana, Mariana and Ady... thanks a lot for sharing a small part of your life, most of you are not (physically) longer here but I will always keep you in a very special place.

Very special thanks to Rober, Bere and Sha for opening not only the doors of your house and family but also those of your heart... sepan que siempre estaremos con ustedes independientemente del tiempo y la distancia, muchas gracias de todo corazón por su amistad e incondicional apoyo! Along the same lines, I would like to thank Stef and Sergiu for being, besides two remarkable people, unconditional friends with whom not only have we shared several laughs but also a shoulder to cry. Guys, you are simply great!

Every house and every building needs a proper foundation... Pa y Ma, quiero agradecerles todo lo que han hecho por mí, los principios y consejos pero sobretodo el saber que ustedes están siempre conmigo... aquellas interminables llamadas telefónicas dominicales se volvieron un ritual para sentirlos más cerca y recargar mis baterías... No saben lo afortunado que me siento por tenerles, muchas gracias por ser como son, siempre un ejemplo a seguir! Xana, Lalo, Lalito y Na, chicos, a todos los llevo siempre en mi corazón y nunca dejo de pensar en ustedes. Los admiro mucho y deseo que todos sus sueños y metas se hagan realidad. Gracias por todos esos momentos inolvidables! Thanks also to Family Guguta for all her help and spiritual support when we mostly need it.

Last but not least, I want to thank one of the persons without whom this might have not been possible, my wife, Carmen... amorcito, mi bolita! Nunca podré agradecerte todo lo que has hecho por mí, por todo lo que hemos vivido, por todo lo que hemos pasado, por tu compresión, apoyo y amor incondicional...sin ti, esto no podría haber terminado de esta manera. Eres mi fuente de inspiración y motivación y lo más valioso que me llevo de Delft. Este es un capítulo más, quedan muchos por delante... siempre junto a ti! Te iubesc foarte mult!

Carlos Manuel López Vázquez
May 2009.

Selected publications

Journal articles

Lopez-Vazquez C. M., Fall C. (2004) Improvement of a gravity oil separator using a designed experiment. *Water Air Soil Pollution*, 157(1-4): 33-52.

Lopez-Vazquez C. M., Song Y.I., Hooijmans C. M., Brdjanovic D., Moussa M. S., Gijzen H., van Loosdrecht M. C. M. (2007) Short-term temperature effects on the anaerobic metabolism of Glycogen Accumulating Organisms. *Biotechnol Bioeng*, 97(3):483-495.

Fall C., **Lopez-Vazquez C.M.**, Jimenez-Moleon M.C., Bâ K.M., Díaz-Delgado C., García-Pulido D., Lucero-Chavez M. (2007) Carwash wastewaters: characteristics, volumes, and treatability by gravity oil separation. *Revista Mexicana de Ingeniería Química*, 6(2): 175-184.

Pinzon Pardo A.L., Brdjanovic D., Moussa M.S., **Lopez-Vazquez C.M.**, Meijer S.C.F., van Straten H.H.A., Janssen A.J.H., Amy G., van Loosdrecht M. C. M. (2007) Application of ASM3 to an oil refinery wastewater treatment plant. *Environ Technol*, 28(11):1273-1284.

Lopez-Vazquez C. M., Hooijmans C. M., Brdjanovic D., Gijzen H., van Loosdrecht M. C. M. (2007) A practical method for the quantification of PAO and GAO populations in activated sludge systems. *Water Environ Res*, 79(13):2487-2498.

Lopez-Vazquez C. M., Brdjanovic D., Hooijmans C. M., Gijzen H., van Loosdrecht M. C. M. (2008) Factors affecting the occurrence of Phosphorus Accumulating Organisms (PAO) and Glycogen Accumulating Organisms (GAO) at full-scale Enhanced Biological Phosphorus Removal (EBPR) wastewater treatment plants. *Water Res*, 42(10-11): 2349-2360.

Lopez-Vazquez C.M., Brdjanovic D., van Loosdrecht M. C. M. (2008) Comment on "Could polyphosphate-accumulating organisms (PAOs) be glycogen–accumulating organisms (GAOs)?" by Zhou et al. Water Res, 42(13): 3561-3562.

Lopez-Vazquez C. M., Hooijmans C. M., Brdjanovic D., Gijzen H., van Loosdrecht M. C. M. (2008) Temperature effects on the aerobic metabolism of Glycogen Accumulating Organisms *Biotechnol Bioeng*, 101(2):295-306.

Lopez-Vazquez C. M., Oehmen A., Zhiguo Y., Hooijmans C. M., Brdjanovic D., Gijzen H., van Loosdrecht M. C. M. (2009) Modelling the competition between Phosphorus- and Glycogen-Accumulating Organisms: temperature, substrate and pH-effects. Water Res, 43(2):450-62.

Lopez-Vazquez C. M., Hooijmans C. M., Brdjanovic D., Gijzen H., van Loosdrecht M. C. M. (2009) Temperature effects on Glycogen Accumulating Organisms. *Water Res* (in press, DOI: 10.1016/j.watres.2009.03.038).

van Loosdrecht M. C. M., Oehmen A, Hooijmans C. M., Brdjanovic D., Gijzen H., Yuan Z., **Lopez-Vazquez C. M.** (2009) Response to comment on "Modelling the PAO-GAO competition: effects of carbon source, pH and temperature" by Dwight Houweling et al. *Water Res* (in press, DOI: doi:10.1016/j.watres.2009.03.043).

Kubare M., **Lopez-Vazquez C.M.,** Brdjanovic D., Amy G., van Loosdrecht M. C. M. (2009) Effect of higher temperature on nitrogen removal in activated sludge wastewater treatment systems (*in preparation*).

Curriculum vitae

Carlos Manuel López Vázquez was born in Toluca, México, on November 30th, 1976. In 1999, he graduated as Civil Engineer from the Faculty of Engineering, Autonomous University of the State of México (UAEMéx), in Toluca, México. He received the Masters degree on Water Sciences, with the specialty on Wastewater Treatment, with distinction (*cum laude*) in 2002 at the Interamerican Center for Water Resources, Faculty of Engineering, at the same university. For outstanding academic achievement, he was awarded different prizes, grants and scholarships including the 'Ingenieros Civiles Asociados (ICA) Scholarship' in 1998 to the best students of Civil Engineering and the 'Ignacio Manuel Altamirano Basilio Prize' in 2001 given to the best MSc. student of the Faculty of Engineering.

In 2003, he was awarded a grant by the National Council for Sciences and Technology (CONACYT) from Mexico to carry out his doctoral studies, which started in 2004 as a joined project between UNESCO-IHE Institute for Water Education and the Environmental Biotechnology group of Delft University of Technology (TUDelft). The results of his PhD research are presented in this thesis.

T - #0081 - 071024 - C20 - 254/178/15 - PB - 9780415558969 - Gloss Lamination